ALL YOU WANT TO KNOW ABOUT THE

MOON

(Q & A)

RAMESH SHISHU, Ph.D.

PRABHAT
PRAKASHAN

Published by
PRABHAT PRAKASHAN PVT. LTD.
4/19 Asaf Ali Road,
New Delhi-110 002 (INDIA)
e-mail: prabhatbooks@gmail.com

ISBN 978-93-5488-663-8
ALL YOU WANT TO KNOW ABOUT THE MOON (Q & A)
by Ramesh Shishu, PhD

Edition
First, 2023

Price
₹ 500 (Rupees Five Hundred Only)

Printed at
Japan Art, Delhi

To

My Students of BITS Pilani

(Pilani, Dubai, Goa, Hyderabad, Mumbai)

Author's Note

This is my third book about the moon. The idea of writing this book emerged during the book launch function of my second book "An Autobiography of the Moon". After a short presentation about the book, there was a "Q & A" session. I realized that "Q & A" is a powerful method of finding quick answers to specific queries. "Traditional Books" are good to get comprehensive knowledge of a subject.

In recent times a renewed interest in the moon as an economic and scientific resource has started a new space race. Many countries are developing technology and plans to put people back on the moon to mine, inhabit, and use it as a launching pad to send spacecraft to other worlds like Mars. This has created an immense interest among people and they are curious to know more about the moon. They have questions and they are looking for answers to their questions. This book about the moon will meet these requirements.

As a first step, with the help of some friends, I prepared a list of "most commonly asked questions about the moon" which should be answered in this book. The list covered the past, present, and future of the moon. I then grouped them into appropriate chapters. The next step was searching for the answers from print and internet media and discussion with knowledgeable researchers and scientists in the field. As expected, the volume of information available in today's electronically connected world is downright scary, if not forbidding. The overwhelming amount of data created confusion in my mind. Several times I found that ploughing through volumes of information was certainly a total turn off. Quite often filtering the information and choosing what to use and what to leave made the process of book writing very slow. But I continued the journey knowing "trees that are slow to grow bear the best fruit." I believe readers will be immensely benefited by saving their time and energy, as they don't have to research the material. I have already done this for them and condensed it in this book. I must admit that the whole process of reading, learning, and writing was great fun and I thoroughly enjoyed it.

In a "Q & A" type book, each Q & A is a short chapter by itself. Therefore, there are duplications. I have tried to minimize such duplications.

For the convenience of the readers, References, wherever required, are given with the answers rather than giving them at the end of the book as conventionally done.

Diagrams are not drawn to scale as it is not possible to draw astronomical diagrams to scale.

I believe that the content of this book is suitable for a general interest audience, who want to know about the moon, as well as those who may have an interest in space and space travel and are looking for some supplemental material.

I hope the readers will find reading this book as interesting as I found documenting it.

❑

Acknowledgements

It has been a great joy for me to research the material and writing this book. There are a few "thank-yous".

I thank Ankur Gupta (Mind Management), Bobby Kumar, R. Rama Raju, Rashmi Varshneya, Roy Gealer, Subhchintan Agarwal, Suresh Mangrulkar, and Vinay Bhatia for assisting me in compiling a list of questions to be answered in this book.

I discussed and corresponded with many researchers, scientists, and writers to understand answers to some of the difficult questions. Many of them granted permissions to use their material (both text as well as figures/images). I thankfully acknowledge their help. They are:

Alexander Novati (NOAA/JPL-Caltech), Ash (timeanddate.com), Brian Koberlein, Camille Carlisle (Sky & Telescope), C. Stuart Hardwick (Science fiction author), Charles A. Wood (Sky & Telescope), Dr. Christopher S. Baired (West Texas A&M University), Claudia Crowley (EarthSky), Damond Benningfield (StarDate Online), Daniel Cummings (Founder - Star in a Star), Elizabeth Sandler (American Association for the Advancement of Sciences), Ethan Siegel (Starts with a Bang), Fraser Cain (Universe Today), G. Jeffrey Taylor (University of Hawaii), Geoff Holt (MMSD Planetarium Wisconsin USA and NASA Airborne Astronomy Ambassador), Ian Garrick-Bethell (University of California, Santa Cruz), Jacinta Bowler (ScienceAlert Pty Ltd), Dr James O'Donoghue (Japan Aerospace

Exploration Agency), Jade Boyd (Rice University), Jason Davis (The Planetary Society), Jatan Mehta (Jatan's Space), Kenneth D. Forbus (Northwestern University), Kevin Stacey (Brown University), Kolvenbach Hendrik (Swiss Federal Institute of Technology), Kristine Spekkens (Royal Military College of Canada and at Queen's University), Lukas Wenzl (Cornell University), Lingzhi sun (University of Hawaii at Manoa), Mamta Patel Nagaraja (Science Mission Directorate, NASA), Mikael Svalgaard, Nancy Chillingworth (Natural History Museum, London), Morgan McFall-Johnsen (Science and Space Reporter at Insider), Dr. Nicholas C. Schmerr (University of Maryland), Rajdeep Dasgupta (Rice University Houston), Ralph Milliken (Brown University), Raluca Rufu (Southwest Research Institute), Randy L. Korotev (Washington University in Saint Louis), Prof. Richard Pogge (Ohio State University, pogge.1@osu.edu), Robert Citron (University of California Davis), Robin Canup (Southwest Research Institute), Rubieyat Bin Ali (Dhaka, Bangladesh), Sarah T. Stewart (University of California at Davis), Shuai Lee (Brown University), Simon J. Lock (California Institute of Technology), Staci L. Tiedeken (NASA's Goddard Space Flight Center), Steffen Thorsen (Timeanddate.com), Stephen M. Elardo (University of Florida), Thomas R. Watters (Smithsonian Institution at Washington), Treehugger.com, Drs. Vanina Lopez de Azarevich (National Scientific and Technical Research Council, Argentina)

My sincere apology if I missed any.

I thank Mahika (Mahi) Inaganti for designing the front and back covers and drawing some figures.

Writing a book is inevitably taxing on the authors' families, who can never be given as much credit as they deserve. To my wife Malti, I am grateful for her patience and understanding and for demonstrating her respect for what it takes to write a book.

❑

Contents

❑

List of Questions

PART I: THE MOON

Chapter 1: The Moon, Size, Mass, and Distance

Q1. What is a moon?

Q2. How many moons are there in our solar system?

Q3. Why do the terrestrial planets (Mercury, Venus, earth, and Mars) have a smaller number of moons in contrast to the Jovian planets (Jupiter, Saturn, Uranus, and Neptune)?

Q4. Why has the earth's moon not been given any name like most other moons in our solar system?

Q5. How big is the moon?

Q6. Is the moon shrinking?

Q7. Why is the moon spherical in shape?

Q8. How far is the moon from the earth? Does the "lunar distance" have any significance in astronomy?

Q9. Is the moon at a constant distance from earth?

Q10. How do we measure the distance between the earth and the moon?

Q11. How do we measure distances of faraway or distant objects in space?

Q12. How did the ancients Greeks calculate the distance to the moon from earth?

Q13. Why is the moon moving away from the earth?

Q14. How close was the moon from the earth when it was formed? How big would the moon have looked at that time?

Q15. Is the rate of departure of the moon away from the earth always been constant?

Q16. What happens to the earth as the moon gets farther away and will it ever go away entirely?

Q17. How do we know the mass of the moon?

Q18. What is the white circle that we see around the moon occasionally?

Chapter 2 - Formation of the Moon

Q19. How did the moon form?

Q20. What is the leading hypothesis/model of the moon's formation?

Q21. What really happened during the moon's formation based on the Giant Impact Model?

Q22. Is the Giant Impact Model of the moon's formation the final answer?

Q23. What other ideas are scientists now working on to explain the moon's formation?

Q24. According to the Giant Impact Model of the moon's formation, the moon is formed from a collision of earth and Theia. What happened to the rest of Theia after the collision?

Q25. Are there any significant contributions made by Theia's collision to proto earth that made the moon?

Q26. How did the early civilizations explain the moon's birth?

Q27. How old is the moon?

Q28. How do scientists estimate the moon's age?

Chapter 3 - Internal Structure of the Moon

Q29. What is the internal structure and composition of the moon?

Q30. How do we know about the internal structure of the moon?

Q31. What is a moonquake?

Q32. How does a moonquake differ from an earthquake?

Q33. What are the different types of moonquakes and what are their causes?

Q34. Does the "moon-shrinking" cause moonquakes?

Q35. Are Moonquakes dangerous to the astronauts and equipment?

Q36. Why do Moonquakes last for a longer time compared to earthquakes?

Q37. Does the moon have volcanic activity now?

Chapter 4 - Lunar Surface and the Moon Mapping

Q38. What are the three main surface features on the moon?

Q39. What are the other important surface features on the moon, other than three main features (dark and light areas, craters, and the moon dust)?

Dark Areas on the Moon (maria)

Q40. What are dark areas on the moon?

Q41. When and how were the maria basins formed?

Q42. How do scientists determine the age of the moon maria?

Q43. Out of two sides of the moon (near side and far side), which side of the moon has more maria basins and why?

Q44. How have maria basins been named?

Q45. What is the largest maria basin and how big is it?

Lighter Areas on the Moon's Surface (Highlands/Mountains)

Q46. What are the lighter areas on the moon's surface?

Q47. How do maria basins differ from highlands?

Q48. Are highlands older than maria basins?

Q49. Does the moon have mountains?

Q50. How were the mountains formed on the moon?

Q51. How are the moon's highlands named?

Q52. What is the highest mountain on the moon and how does it compare with earth's highest mountain?

Q53. Which regions of the moon have the highest and the lowest elevations?

Craters

Q54. Why does the moon have craters?

Q55. How are craters made on the moon?

Q56. What are some of the important features of the moon craters?

Q57. How are the moon's craters named?

Q58. How many craters are on the moon?

Q59. Why has the moon's far side more craters than near side?

Q60. Why are the highlands more cratered than the maria?

Q61. Which is the largest crater in the moon?

Q62. Why does the moon have so many craters while earth has so few?

Q63. Why are the moon craters larger in the "near side" of the moon than in the "far side"?

Rocks

Q64. How were rocks formed on the moon?

Q65. Why does the moon have big rocks?

Q66. What kinds of rocks are there on the moon?

Q67. What are breccias and how did they form on the moon?

Q68. How old are the moon rocks?

Q69. Why are the rocks on the moon older than those on earth?

Moon Dust (Regolith)

Q70. What is lunar dust (regolith) made of?

Q71. What other terms are being used in reference to lunar regolith?

Q72. Is the moon dust dangerous?

Q73. How can lunar regolith be used for human purposes?

Q74. What are the moon dust fountains?

Other Features on the Moon's Surface

- **Lunar Rays**

Q75. What are the lunar rays?

Q76. How did lunar rays form?

- **Lunar Valleys / Rilles**

Q77. What are lunar valleys/rilles and how are they formed?

Q78. What are the different types of rilles?

- **Lunar Lava Tubes**

Q79. What are lunar lava tubes and how are they formed?

Q80. How do lava tubes differ from tubes found on earth?

Q81. How can lunar lava tubes be used for human purposes?

Q82. Are lunar lava tubes intact and stable and not prone to future collapse and asteroidal bombardment?

- **Ridges and Cracks / Dorsa**

Q83. What are wrinkle ridges on the moon?

Q84. How are wrinkle ridges formed on the moon?

Q85. How are wrinkle ridges named?

- **Lunar Swirl (Albedo)**

Q86. What are lunar swirls?

Q87. How are lunar swirls formed?

Q88. Which theory of lunar swirls formation is supported by recent measurements/observations?

- **Moon Mapping**

Q89. How has the moon been mapped through time?

Chapter 5 - Lunar Atmosphere

Q90. Is there an atmosphere on the moon?

Q91. Why doesn't the moon have an atmosphere?

Q92. What is the moon's thin atmosphere made of and what are the sources?

Q93. What are the differences between the atmospheres of earth and the moon?

Q94. Did the ancient moon once have an appreciable atmosphere?

Q95. What are the consequences of the moon's non-existent or a thin atmosphere?

Q96. Why is there so much temperature variation on the moon?

Q97. How do astronauts survive huge temperature variation on the moon?

Q98. Does the moon have weather?

Chapter 6- Lunar Gravity

Q99. Does the moon have gravity?

Q100. Why is the moon's gravity so uneven?

Q101. Is low gravity of the moon a health hazard?

Q102. How does the gravitational force of the moon affect earth?

Chapter 7 - Magnetic Field of the Moon

Q103. Does the moon have a magnetic field?

Q104. Why does earth have a magnetic field, whereas the moon does not?

Q105. Is varying magnetic field strengths of the moon related to its geology?

Q106. Was there a strong magnetic field on the moon earlier?

Q107. What is the mystery of the moon's lost magnetism?

Q108. Is it a confirmed conclusion that the moon once had a Magnetosphere?

Q109. Do magnets work on the moon?

Q110. Is there a risk of astronauts getting electric shocks on the moon's surface?

Chapter 8 - Chemicals on the Moon

Q111. What chemical elements are found on the moon?

Q112. What elements are lacking in the moon rocks?

Q113. What are the primary minerals on the moon?

Q114. Is there a relationship between lunar mineral distributions and lunar geology?

Q115. Why is the moon depleted in volatile elements?

Q116. Which chemicals and mineralogical resources of the moon can be used?

Chapter 9 - Helium-3 on the Moon

Q117. What is Helium-3 (He3) and why is it so important?

Q118. How did the moon acquire Helium-3?

Q119. How much Helium-3 (He3) is there on the moon?

Q120. Why is Helium-3 a better nuclear fuel?

Q121. What are the challenges in using the moon's helium-3 for generating power?

Chapter 10 - Water on the Moon

Q122. Is water available on the moon?

Q123. How and when was the presence of water on the moon confirmed?

Q124. Are there more confirmations of water on the moon after the Chandrayaan-1 mission?

Q125. How can water exist on the moon when it has no atmosphere and daytime temperatures reach 120 degrees Celsius (250 degrees Fahrenheit)?

Q126. Where did water on the moon come from?

Q127. How much water is estimated on the moon?

Q128. How will future missions study water on the moon?

Q129. What are the challenges in exploring the moon's water ice directly?

PART II: MOON IN MOTION

Chapter 11 - Moon in Motion: Orbital and Rotational

Moon's Orbital Motion

Q130. Why does the moon orbit around earth?

Q131. Why do objects in space follow elliptical orbits?

Q132. The sun is larger than the earth and has more gravitational force. Then why does the moon not orbit around the sun instead of the earth?

Q133. What are some of the moon's orbital details?

Q134. Why does the moon's apparent size change across its orbit, sometimes looking bigger and sometimes smaller (not to confuse with phases of the moon)?

Q135. Why do the planets and their moons all rotate / orbit in the same plane?

Q136. Why does the moon appear higher in the winter night sky and lower in the summer nights?

Q137. Earth's gravity pulls the moon then why doesn't the moon fall into the earth and crash?

Moon's Rotational Motion

Q138. How long is a day on the moon?

Q139. Why do we see only one side of the moon?

Q140. What are the sidereal and synodic months and why are they different?

Q141. Why does the moon rise and set at different times and in different locations?

Q142. What is the earth – moon Tango?

Q143. Why does the moon wobble?

Q144. What is the moon's apsidal/orbital precession?

Chapter 12 – Near Side and Far Side of the Moon

Q145. What are the near side and far side of the moon?

Q146. The far side of the moon is also referred or described as "dark side". Does it mean it never receives sunlight?

Q147. Have we seen/visited the far side of the moon?

Q148. What are the major differences between the near side and far side of the moon?

Q149. Why are the two sides of the moon (near side and far side) different from each other?

Q150. How can the far side of the moon be beneficial to glimpse the beginnings of the universe?

Chapter 13 - Moon Phases

Q151. What causes moonlight?

Q152. Why is the moon so bright?

Q153. Is there any difference in appearance of the moon viewed in the southern hemisphere (say Australia) and northern hemisphere (say New York)?

Q154. Why does the moon seem to change its shape every night?

Q155. How were the moon's phases named?

Q156. Are moon phases the same everywhere on earth? In other words, is it the full-moon in Sydney at the same time as the full-moon in Delhi, London, and New York?

Q157. What are the many names given to full moons?

Q158. What is a blue moon?

Q159. On a full moon, the earth is between the sun and the moon. Then how does the sunlight get to the moon?

Q160. How during a thin crescent moon can one sometimes see the outline of the rest of the moon?

Q161. How can we sometimes see the moon during the day?

Q162. What is a Super moon and a Micro moon?

Q163. Do the moon phases affect the rain on the earth?

Q164. Does the full moon affect animals' (especially wolves) or people's behaviour?

Q165. Why does the moon seem yellow, orange, or red when it is near the horizon?

Q166. Just like we see the moon's phases from earth, are earth's phases also visible from the moon?

Chapter 14 - Celestial Shadows- Transit, Eclipse, and Occultation

Q167. What is an eclipse?

Q168. What are umbra and penumbra?

Q169. What are lunar and solar eclipses?

Q170. What are the different types of lunar eclipses?

Q171. Why does the moon look red during a total lunar eclipse?

Q172. What are the different forms of solar eclipses?

Q173. What is the diamond ring effect in a solar eclipse?

Q174. The sun is much larger than the moon, then how do we get a total solar eclipse?

Chapter 15 - Tides

PART III: EXPLORATION

Chapter 16 - Science of Space Travel

Q195. What are Kepler's laws of orbital motion?

Q196. What is Tsiolkovsky's rocket equation?

Q197. How is Tsiolkovsky's rocket equation used?

Q198. What conclusions can be drawn from the Tsiolkovsky's rocket equation?

Q199. What is colloquially known as the "tyranny of the rocket equation"?

Q200. How are the science laws used in space travel?

Q201. How do satellites stay in orbit?

Chapter 17 - Space Race Between the USA and the USSR (Soviet Union)

Q202. What started the space race between the USA and Soviet Union (USSR) in the1950s?

Q203. What were some of the initial events of the space race between the USA and USSR? (Up to the first Soviet and the first American orbited the earth)

Q204. How did the space race between the USA and USSR benefit the world?

Chapter 18 - Journey to the Moon and Lunar Exploration

Q205. Who discovered/studied the moon before the Apollo mission?

Q206. Why is space exploration important?

Q207. Why do we want to go back to the moon?

Q208. How many people have been to space to date?

Q209. How many countries have been to the moon (on its surface or in its orbit)?

Q210. Why have humans never been back to the moon after 1972?

Q211. What various spacecraft/vehicles have been used so far for the moon exploration, and what did we learn from them?

Q212. Which important moon exploration activities are being focused on now by global space agencies?

Chapter 19 - USA's Lunar Exploration Programme

Q213. What were the various stages of USA's lunar exploration programme?

Q214. What were the goals of various NASA's probes (robots) used in lunar exploration activities?

Q215. What was NASA's project Mercury?

Q216. What was NASA's project Gemini?

Q217. What was NASA's project Apollo?

Q218. What spacecraft were used for the Apollo project?

Q219. How did Apollo astronauts land on the moon and return to the earth?

Q220. What are some of the milestones of the Apollo project?

Q221. What are NASA's lunar/space exploration activities after the Apollo project?

Q222. What is NASA's Artemis project?

Q223. What are the main phases of the Artemis project?

Q224. How is NASA's Artemis project different from the Apollo project?

Q225. What are NASA's Artemis's planned and future missions?

Q226. What is Artemis accords?

Q227. What is NASA's VIPER lunar rover?

Q228. What are NASA's plans for a lunar base camp?

Chapter 20 - USSR's/ Russia's Lunar Exploration Programme

Q229. What were the Soviet Union's lunar/space exploration missions during 1959-1979?

Q230. What were the Soviet Union's "the first historical space activities"?

Q231. What are Russia's future lunar exploration projects?

Q232. What are Russia's and China's joint space/lunar exploration projects?

Chapter 21 - China's Space and Lunar Exploration Programme

Q233. When did China start its space programme?

Q234. What is China's Shenzhou programme?

Q235. What were China's space laboratories Tiangong-1 and Tiangong-2?

Q236. What is China's Tiangong space station?

Q237. What are some of the salient features of China's space station?

Q238. What are Chinese lunar exploration activities?

Q239. What is the Chinese next-generation reusable crewed spacecraft?

Chapter 22- India's Space and Lunar Exploration Programme

Q240. When did India start the space exploration programme and what are its major activities?

Q241. What is India's lunar exploration programme?

Q242. What are some of the salient features of Chandrayaan-1?

Q243. What are some of the salient features of Chandrayaan-2?

Q244. What did Chandrayaan-2 discover?

Q245. Why are we interested in the moon-landings on moon's south pole?

Q246. What are ISRO's past and future space/lunar exploration missions?

Chapter 23 - Europe's Lunar Exploration Programme

Q247. Which European agency is involved in the space/lunar exploration activities?

Q248. What are the European Space Agency's lunar exploration missions?

Q249. What was the European Space Agency's first lunar probe?

Q250. Why does the European Space Agency's SMART-1 represent a breakthrough in planetary exploration?

Q251. What is European Space Agency's SpaceBok robot?

Chapter 24 – Japan's Lunar Exploration Programme

Q252. What was Japan's Hiten programme?

Q253. What is Japan's Kaguya lunar exploration mission?

Q254. What are Japan's lunar exploration projects?

Chapter 25 – Lunar Exploration Programs of Other Countries

(Australia, Canada, Emirates, Israel, South Korea, Turkey, and UK's Spacebit)

Q255. What is Australia's lunar mission?

Q256. What is Canada's role in lunar exploration activities?

Q257. What is Emirates' lunar mission?

Q258. What are Israel's lunar exploration activities?

Q259. What is South Korea's lunar exploration programme?

Q260. What is Turkey's lunar exploration programme?

Q261. What is Spacebit's (UK) walking robot Asagumo?

PART IV: OTHER TOPICS

Chapter 26 – Other Topics

- **Legal Issues**

Q262. What is the Space Law?

Q263. What are the five international treaties governing outer space?

Q264. What are the five sets of principles governing outer space?

Q265. Can you buy a part of the moon?

- **What if the Moon was not there?**

Q266. What would have happened if we did not have the moon?

- **Health Issues**

Q267. How does space/lunar exploration affect human health?

- **Moon Landing Hoax**

Q268. Why do some people still think that the moon landings were a hoax and what are NASA's defence arguments?

- **Moon's Effects on Life on earth**

Q269. How does the moon affect life on earth?

- **Moon in ancient civilization/culture**

Q270. Why is the moon so important in many cultures?

- **Russia - Ukraine War: Impact on Space Programme**

Q271. Will Russia's war in Ukraine hinder space research and exploration?

❑

PART I: THE MOON

The moon is a friend for the lonesome to talk to.

– Carl Sandburg

CHAPTER 1

The Moon, Size, Mass, and Distance

Q1. What is a moon?

A1. The astronomical community has never got around to making a formal definition of a moon. So, as it stands, a moon is simply any natural celestial body that orbits a planet or something else that is not a star. Besides planets, moons can orbit dwarf planets, large asteroids, and other bodies. Since objects that orbit other objects are called satellites, therefore moons are also known as natural satellites

All the planets in our solar system, except Mercury and Venus, have moons. Pluto and some other dwarf planets and many asteroids also have moons.

Moons come in many shapes, sizes, and types. As of now there is no minimum size required for naming a body a moon. Known moons in our solar system range in size from Jupiter's giant moon Ganymede, about 1.5 times the earth's moon (larger than the planet Mercury at 5,268 km or 3,273 miles across) down to small bodies less than 10 km (6 miles) across.

Most planetary moons were probably formed from the discs of gas and dust circulating around planets in the early solar system. Some are "captured" objects that formed elsewhere and fell into orbit around larger bodies due to gravitational attraction. The earth's moon was likely formed after a Mars-sized body collided with earth. (More on this later)

Q2. How many moons are there in our solar system?

A2. It is a difficult question to answer as the astronomical community has never made a formal definition of a moon.

Keeping that in mind, if we call moons most people are familiar with as "traditional" moons (moons of Earth, Mars, Jupiter, Saturn, Uranus, Neptune, and Pluto) then as on May 2023, this count stands at 289 (See table below).

This includes two recent discoveries:

1. A team of astronomers led by Scott Sheppard of the Carnegie Institution has found and confirmed (Feb 2023) 12 new moons orbiting Jupiter, bringing the total of moons at the giant planet to 95.
2. In May of 2023, the International Astronomical Union is set to recognize 62 additional moons of Saturn based on a batch of objects discovered by astronomers. The small objects will give Saturn 145 moons (which was 83 earlier). The details of this recent discovery are given in References number 3 and 4 given below.

In addition to this, astronomers have documented another 464 moons orbiting smaller objects, such as asteroids, Kuiper Belt and Trans-Neptunian Objects. That makes a total count of moons as 753.

There are likely thousands more moons awaiting discovery in our solar system.

Terrestrial Planet	Number of Moons
Mercury	0
Venus	0

earth	1
Mars	2
Jovian Planet	
Jupiter	95
Saturn	145
Uranus	27
Neptune	14
Dwarf Planet	**Number of Moons**
Pluto	5
TOTAL	**289**

moons orbiting smaller objects, such as asteroids, Kuiper Belt and Trans-Neptunian Objects (As per last update of April 14, 2023, by JPL Solar System Dynamics) 464.

That makes a total of 753.

Source: NASA (updated by the author of this book)

References

1. *Davis, Phil. Cerney, Steve, Moons. Solar System Exploration. NASA's Science Mission Directorate. Last update April 28, 2023. https://solarsystem.nasa.gov/moons/overview*
2. *Earth & Planets Laboratory. Carnegie Science. Jupiter back on top as "moon making". February 10, 2023. https://carnegiescience.edu/jupiter-back-top-moon-king*
3. *O'Callaghan, Jonathan. With 62 Newly Discovered Moons, Saturn Knocks Jupiter Off Its Pedestal. New York Times. Published May 12, 2023Updated May 15, 2023. https://www.nytimes.com/2023/05/12/science/saturn-moons-jupiter.html*
4. *Sullivan, Will. Saturn May Have Just Won the 'Moon Race' With 62 More Discovered. Smithsonian Magazine (Smart News). May 16, 2023. https://www.smithsonianmag.com/smart-news/saturn-may-have-just-won-the-moon-race-with-62-more-*

discovered-180982180/#:~ :text=But%20over%20 the%20last%20couple, Columbia%20(UBC) %20in%20 Canada.

Q3. Why do the terrestrial planets (Mercury, Venus, earth, and Mars) have a smaller number of moons in contrast to the Jovian planets (Jupiter, Saturn, Uranus, and Neptune)?

A3. The number of moons for each planet depends on three factors:

1. Size / Mass of the planet,
2. Planet's proximity to the sun, and
3. Nearness of the planet to the asteroid or Kuiper belt (a rocky asteroid belt).

The primary reason is the gravitational pull. The bigger (more mass) the planet, stronger the gravitational pull. The planets in our solar system that have more moons are the large planets Jupiter, Saturn, Uranus, and Neptune. They are so big that they can collect many moons. Mercury and Venus have too weak gravitational pull to grab a passing object and keep it in its orbit. See the list below:

Surface Gravity of the Planets and the Sun

Rank	Name	Surface Gravity (metrr pr. square second)
1	Sun	274
2	Jupiter	24.92
3	Neptune	11.15
4	Saturn	10.44
5	Earth	9.798
6	Uranus	8.87
7	Venus	8.87
8	Mars	3.71
9	Mercury	3.7
10	moon	1.62
11	Pluto	0.58

Source: NASA

The sun, having more mass, absorbed most of the material that was leftover. The farther from the sun you are, the easier it is to capture or maintain a satellite.

Most planets form when the small chunks of rocks in the solar system start to come together and form larger and larger planets. If these planets were close to asteroid belts, you may get more moons because those asteroids will be captured by the planets. Pluto has five moons because it was near Kuiper belt, a rocky asteroid belt. Just because it was near those minuscule rocky leftovers, so it could grab some of them.

Q4. Why has earth's moon not been given any name like most other moons in our solar system?

A4. Most of the moons of our solar system have names taken directly from Greek and Roman mythology or Shakespearean characters. But earth's moon has not been given any name. It is simply known as "the moon", with capital "M".

There is a good reason for this.

People did not know about the existence of other moons till Galileo Galilei discovered four moons of Jupiter in the year 1610. When humanity first learned of other moons orbiting the planets in our solar system, one of the primary reasons they were given names was to differentiate them from the earth's moon. The naming of moons was one of the first things the International Astronomical Union (IAU), the entity responsible for naming every celestial body in the night sky, did when it was formed in 1919. For earth's moon, IAU settled on "the moon" instead of any other names. They stated that since the moon was the first known moon, and popularly called like that for millennia in many languages, therefore it was easier to simply make it official rather than introducing a new name.

So, earth's moon is known as "the moon".

Regarding the etymology of the word moon, it can be traced to the word mōna, an Old English word from medieval times. Mōna shares its origins with the Latin words metri, which means to measure, and mensis, which means month. So, the moon is called "the moon" because it is used to measure the months.

Q5. How big is the moon?

A5. Diameter: The diameter of the moon is 3,475 kilometres (2,159 miles). It is less than a third the diameter of the earth and is the fifth largest moon in our solar system. Its circumference is 10,920 kilometres (6,784 mi).

Volume: The volume of the moon is about 221.9 billion cubic kilometres (5.28 billion cubic miles) which is only 2 per cent the volume of the earth. The volume of the earth is 1 trillion cubic kilometres. That means 49 moons would fit inside the earth.

Surface Area: The moon's surface area is about 38 million square kilometres (14.6 million square miles), which is only 7.4% of the earth, 3.85 times the area of USA, 11.5 times the area of India and 85 per cent of the continent Asia.

Mass and density: The moon's mass is 7.35 x 10^22 (twenty zeros after 735) kg, about 1.2 per cent of earth's mass. In other words, earth weighs 81 times more than the moon. The moon's density is 3.34 grams per cubic centimeter (3.34 g/cm3), which is about 60 per cent of earth's density. The moon is the second densest moon in the solar system; Jupiter's moon "Io" is denser, with a density of 3.53 g/cm3.

Q6. Is the moon shrinking?

A6. Yes, the moon is shrinking. Astronomers have spotted wrinkles all over the lunar surface. The contraction marks were discovered by US scientists of Smithsonian Institution in Washington DC. The team examined thousands of photographs of the moon's surface taken by NASA's Lunar Reconnaissance Orbiter (LRO). Some of the wrinkles are several miles long and rise tens of metres above the surface. Researchers believe they arise from the moon decreasing in size by around 200 metres across its diameter in the last billion years.

The prospect of a shrinking moon is not new to planetary experts. When the moon formed it had a hot core which caused it first to expand and then contract as it cooled down. The latest findings suggest the moon could still be cooling, a process that causes the surface to compress and form the wrinkle-like features, known as lobate scarps.

Similar markings were photographed on the moon during the Apollo missions. Scientists will now compare those pictures with the latest images to see if anything has changed in the past 50 years.

Q7. Why is the moon spherical in shape?

A7. To the eye, the moon appears round, and it is natural to assume that it is spherical in shape. But the fact is that the moon is not perfectly round. It is little bit lemon-shaped, known as oblate spheroid (meaning it has the shape of a ball that is slightly flattened) in which the equatorial diameter is greater than the polar diameter. Such shapes are also called ellipsoids. The moon's diameter from side to side is 4.34 km more than its diameter from pole to pole.

The moon and other celestial bodies are approximately spherical due to gravitational pull during their formation. Gravitational pull is a force that points inwards from all directions towards a centre. Mass and speed of rotation are important factors to make the final shape. Less massive objects, such as asteroids, comets, and smaller moons have less gravity and can be non-spherical. When a body becomes sufficiently massive, the shape adjusts itself till all parts of the body are roughly equidistant from the centre and no part of the body is gravitationally unbalanced. Only a spherical shape can accomplish this. Rotation makes a body flattened at the pole. Jupiter spins more rapidly and is flattened more. Stars are similar. Our sun rotates slowly, so it is almost a perfect sphere, but there are stars that spin very fast. Any faster and it would tear itself apart from centripetal forces.

The moon has a weird bulge on both the near side (the side of the moon that faces earth) and far side (the side away from us), giving it a shape like a lemon. Also, near side is slightly larger than the far side. This makes it slightly like the shape of typical bird egg that is larger on one end than on the other.

So, the moon is not exactly spherical, it is more like a lemon. The deviation is small, but real.

According to astrophysicists, Ian Garrick-Bethell along with his colleagues and Maria Zuber and her colleagues, the moon's own spin and the tidal tug of the earth created a "lemon-shaped" moon. During the early stage of moon formation, when it was still superhot, the earth's gravitational (tidal) forces moulded its shape. Later, those tides warped the outside of the moon while it was cooling, causing its crust to expand outward in certain places and it froze in that warped shape. The team estimated that the bulges formed gradually in the first 200 million years of the moon's existence, due to the combination of forces, rotational and tidal. The moon is a bit lemon-shaped with the long axis of the lemon pointing at the earth.

References

1. *Gates, Sara. Scientists Finally Know Why the Moon is Shaped Like a Lemon. Science. July 31, 2014.* https://www.huffpost.com/entry/moon-lemon-shape-mystery-tidal-forces_n_5635386

2. *Hitchings, Lauren. Leaving earth Made the Moon Lemon-Shaped. New Scientist, SPACE. July 30, 2014* https://www.newscientist.com/article/dn25976-leaving-earth-made-the-moon-lemon-shaped/

3. *Garrick-Bethell, I., Perera, V., Nimmo, F. et al. The Tidal–Rotational Shape of the moon and Evidence for Polar Wander. Nature 512, 181–184 (2014).* https://doi.org/10.1038/nature13639

Q8. How far is the Moon from the earth? Does lunar distance have any significance in astronomy?

A8. The Moon is an average of 384,400 kilometres (238,855 miles) away from the earth. That means 30 earth-sized planets could fit in between earth and the moon. This is a very short distance in terms of space terminology. Based on the current speeds, it takes about 3 days for a spacecraft to reach the moon.

Lunar distance (LD) is a unit of measure in astronomy commonly used to express the distance to near-earth objects. Light from the moon reaches in 1.28 seconds to the earth. Compare this to sun's light which reaches in 8 minutes and 20 seconds.

For larger distances in space, the units used are Astronomical unit (AU), the light-year (ly), and parsecs(pc).

The Astronomical Unit (AU) is the basic unit for measuring the separation between bodies in the solar system. One AU represents the mean distance between the earth and the sun which is 149,597,871 km (92,955,807 miles).

The Light-Year (ly) is the distance light travels in a year which is equal to 9,460,730,472,581 kilometres (approximately 5,878,625,400,000 miles). This is a popular measure of intergalactic distances.

The Parsec (pc) or "parallax second" is a unit of length used to measure the astronomically large distances between objects beyond our Solar System. One parsec is the distance at which one Astronomical Unit (AU) subtends an angle of one arcsecond. One parsec equals 3.26 light-years or 206,000 astronomical units (AU), i.e., 30.9 trillion kilometres (19.2 trillion miles).

Even parsecs prove inadequate for measuring distances within the galaxy or intergalactic distances. Astrophysicists frequently express these in kiloparsecs and megaparsecs, which equal 1,000 and 1 million parsecs, respectively. For example, the centre of the Milky Way galaxy from earth is about 8 kiloparsecs distant.

Lunar distance is an important astronomical datum as it can be measured very precisely. This is used for testing gravitational theories such as general relativity and for refining other astronomical values such as earth's mass, radius, rotation, lunar radius, mass of the sun, and the distance to the sun.

Notes: Millimeter-precision measurements of the lunar distance are made by measuring the time taken for light to travel between LiDAR (Light Detection and Ranging) stations on the earth and retroreflectors placed on the moon. LiDAR uses a pulsed laser to calculate an object's (like the moon, an airplane, etc.) variable distances from the earth surface.

Q9. Is the moon at a constant distance from earth?

A9. The moon is not always at the same distance away from the earth. That is why we always say that the average distance of the moon from earth is 384,400 kilometres (238,855 miles). The moon's orbit around earth is not a perfect circle. The actual distance varies over the course of the orbit from 356,500 km (221,500 mi) at the closest point known as perigee to 406,700 km (252700 mi) at the farthest known as apogee, resulting in a differential range of 50,200 km (31,200 mi). The instantaneous lunar distance is constantly changing. In fact, the true distance between the moon and earth can change as quickly as 75 metres per second, or more than 1,000 km (620 mi) in just 6 hours, mainly due to its non-circular orbit.

Q10. How do we measure the distance between the earth and the moon?

A10. Astronomers use different methods to determine the distances to various objects in space. The specific method which is used depends on how far away the object is. For short distances, like distance between the moon and the earth, the Radar technique is used. Scientists measure the distance between earth and the moon with the help of a radar, using known value of speed of light and by timing a pulse of radio waves to the moon and back. Retroreflectors left by Apollo astronauts reflect a laser beam shot from earth and allow scientists to keep close track on the lunar orbit. The measurements are ongoing and involve multiple laser facilities. The instantaneous precision of the Lunar Laser Ranging experiments can exceed sub-millimeter resolution and is the most reliable method of determining the lunar distance, to date.

Q11. How do we measure distances of faraway objects in space?

A11. This is going to be complicated. For simplicity, we will mention here the names and principle involved of the techniques.

Measuring distances to far away objects in the universe is not easy. Astronomers use different methods depending on how far away the object is. All the methods are wonderful combinations of science and mathematics.

The following techniques are used:

1. Radar - measuring short distances in our solar system

The technique is based on measuring the time taken by a reflected light signal and multiplying by the velocity of light 300,000 km/sec (186,000 miles/sec).

2. Parallax - measuring distances to nearby stars

In astronomy, parallax is the apparent motion of nearby objects against their background caused by the earth's travel around the sun. Astronomers estimate the distance of nearby stars in space by using this parallax phenomenon. Because closer stars seem to move more than distant ones, the amount of apparent motion allows astronomers to determine their distances by measuring the change in the angle of observation as it appears from earth.

3. Cepheids - measuring distances in our Galaxy and to nearby galaxies

Cepheids are especially bright stars and are visible in galaxies that are tens of millions of light-years away. These stars brighten and dim periodically, pulsating in and out like beating hearts. This behaviour allows them to be used as cosmic yardsticks out to distances of a few tens of millions of light-years. There is a direct relationship between the length of a Cepheid's pulsation and its true brightness which in turn is related to its distance. Astronomers first use the parallax method to get the distances to some of the closer Cepheids. This allows them to calibrate a Cepheid's true brightness, which then can be used to calculate its distance.

4. Supernovae - measuring distances to other galaxies

For more-distant galaxies, astronomers rely on the exploding stars known as supernovae. The unique characteristics and enormous brightness of a certain type of supernova, the explosion which can occur at the end of the main sequence life of a massive star, can be used to determine distances beyond the reach of the previous methods. Like Cepheids, the rate at which a certain class of supernovae brighten, and fade reveals their true brightness, which can then be used to calculate their distance. A good calibration using parallax and Cepheids is a must without which the technique would not work.

5. Redshift and Hubble's Law - measuring distances to objects far, far away

For extremely far objects (beyond about 1 billion light-years), none of the above methods work. For such distances, scientists use observations in conjunction with Edwin Hubble' theory of expanding universe. In 1929, Edwin Hubble announced that the universe was expanding with all the galaxies moving away from each other. This phenomenon was observed as a redshift of a galaxy's spectrum. This redshift appeared to be larger for faint, presumably further, galaxies. Hence, the farther a galaxy, the faster it is receding from earth.

The velocity of a galaxy could be expressed mathematically as

$$v = H \times d$$

where v is the galaxy's radial outward velocity, d is the galaxy's distance from earth, and H is Hubble constant, the rate at which the universe is expanding.

To determine an object's distance, we only need to know its velocity. By taking the spectrum of a distant object, such as a galaxy, astronomers can see a shift in the lines of its spectrum and from this shift determine its velocity. Putting this velocity into the Hubble equation, they determine the distance.

Q12. How did the ancients Greeks calculate the distance to the moon from earth?

A12. The development of astronomy by the Greeks is a major phase in the history of astronomy. Nowadays scientists measure the distance between earth and the moon with the help of radar, using known value of speed of light and by timing a pulse of radio waves to the moon and back. Greek astronomers were able to estimate radius of the earth and the distance of the moon and sun with reasonable accuracy 22 centuries ago with no radar, no radio time signals, and no telescopes.

Aristarchus of Samos (320–250 BC) used lunar eclipse to find the size of the moon and then using geometry, estimated how far it is from the earth. You can also do that. It is a simple matter of tracking and timing how long it takes the earth's shadow to cross over the moon. Ancient Greeks knew that the moon travels around the earth at a constant speed of about 29 days per revolution. From earlier estimation they also knew the diameter of the earth to be about 12,875 km or 8,000 miles. By tracking the movement of the earth's shadow across the moon, Greek astronomers found that the earth's shadow was roughly 2.5 times the apparent size of the moon and lasted roughly three hours from the first to last signs of the shadow. From these measurements, it was simple geometry that allowed Aristarchus (c. 270 BC) to determined that the moon was around 60 earth radii away (about 386,243 km or 240,000 miles). This is quite close to the currently accepted figure of 60.3.

Q13. Why is the moon moving away from the earth?

A13. The moon is moving away from the earth. Scientists refer to this as "lunar retreat." Nearly 300 years ago, astronomer Edmond Halley was the first one who suspected that the moon was receding after studying records of ancient eclipses. This finally got confirmed in the 1970s as detected by the Lunar Laser Ranging (LLR) Experiment which showed that the moon is moving away at a rate of 3.8 cm (1.5 inches) per year. LLR is the practice of measuring the distance between the surfaces of the earth and the moon using a laser rangefinder (also known as a laser telemeter) that uses a laser beam to determine the distance to an object. The distance between earth and the moon can be calculated from the round-trip time of laser light pulses travelling at the speed of light, which are reflected back to earth by the moon's surface or by one of five retroreflectors installed on the moon during the Apollo programme (11, 14, and 15) and Lunokhod missions.

The moon is driven away by the effect of its gravity on the rotating earth. Tides raised in the oceans cause drag and thus slow the earth's spin-rate. The resulting loss of angular momentum is compensated for by the moon speeding up, and thus moving further away. (More on this in the chapter about "Tides")

Q14. How close was the moon from earth when it was formed? How big would the moon have looked at that time?

A14. It is not an easy question to answer as it depends primarily on the unknown details of the impact of Theia and the early earth responsible for the moon's formation and how the hot material behaved in space. But simulations suggest it was very close to earth, about 3-5 times the radius of the earth, or about 20 to 30 thousand kilometres (12,400 to 18,600 miles). Scientists have given a number 24,000 Km (15,000 miles). The moon is currently 384,400 km (238,855 miles) or 60 earth radii away from earth, which means that now the moon is about fifteen times further away than it was when it first formed.

Its apparent size was about 15 times bigger than what it appears now (again an estimate based on simulation and modelling).

Q15. Has the rate of departure of the moon away from the earth always been constant?

A15. The exact rate of the moon's movement away from earth has varied a lot over time. It depends both on the distance between the earth and the moon, and the shape of the earth. The details of continents and oceans moving around on earth change the rate, which make it an extremely hard thing to estimate.

An interesting animation (using the data of V.L.L. de Azarevich, and M. Bshows Azarevich) by James O'Donoghue, a planetary scientist at the Japan Aerospace Exploration Agency (JAXA), has shown moon's 4.5-billion-year journey, from birth till now. This has been discussed by Morgan McFall-Johnsen in her paper in Insider as follows:

"The sequence starts with the moon's current position and follows it back in time to its birth, tracking its distance from earth, apparent size relative to our planet, and the speed of its retreat over time... The pace of this motion has not always been constant. According to O'Donoghue's animation, the moon started out moving away at 20.8 centimeters (8.2 inches) per year, and its drift has fluctuated from between 0.13 centimeters (0.05 inches) per year to 27.8 centimeters (10.9 inches) per year, ...compare these with the current drift of 3.8 centimeters (1.5 inches). Much of the variation in the moon's pace of movement comes from lunar meteorite impacts and major geological changes on earth".

References

1. *López de Azarevich, Vanina Lucrecia and Azarevich, Miguel. (2017). Lunar Recession Encoded in Tidal Rhythmites: A Selective Overview with Examples from Argentina. Geo-Marine Letters. 37.*10.1007/s00367-017-0500-z.
2. *Johnsen, Morgan McFall. The moon has been drifting away from Earth for 4.5 billion years. A stunning animation shows how far it has gone. Insider. Oct 1, 2019.* https://www.businessinsider.com/video-moon-drifts-away-earth-4-billion-years-2019-9#: ~ :text=Scientists%20refer%20to%20this%20as,(10.9%20inches)%20%20year.

Q16. What happens to the earth as the moon gets farther away and will it ever go away entirely?

A16. According to the conservation of momentum, the earth slows down as the moon gets farther away. This means that the length of the day increases - more hours in a day. However, this increased day length will be a small fraction of a second in our lifetime.

The moon will never go away though. After billions of years (the precise number is hard to estimate), the moon will get to its final end state where it orbits more slowly. At that state, the moon would take about 47 days to orbit the earth (currently, it takes a little over 27 days). earth's spin would also have slowed to one rotation every 47 days. One side of the earth will see the moon fully up in the sky while the other side of the earth will never see the moon. This state is called tidally locked. By this time, the sun is going to turn into a Red Giant in about 5 billion years, and many scientists believe that it will consume the earth in the process. So, the earth and moon will be destroyed completely long before they achieve a tidally locked state, and the human race will be extinct.

Q17. How do we know the mass of the moon?

A17. The mass on any heavenly body can be estimated through its gravity using the following relation:

$$M = (g \times R^2)/G$$

Considering the example of the earth, M is the mass of the earth, G is gravitational constant, R is earth's radius, g is acceleration due to gravity on earth.

The earth is an easier case as we can measure its acceleration due to gravity simply by dropping an object from a height to its surface. Since humans have visited the moon, its gravity is also known by the hammer and feather dropping experiment done on the surface of the moon and by several measurements done by instruments and American lunar science mission Gravity Recovery and Interior Laboratory (GRAIL).

But how about the mass of other planets since no human ever visited them and measured their gravity on the spot. We must usually resort to other methods. The other technique to find the gravity of any heavenly body is to observe another body orbiting or passing close to that heavenly body and see how its path is affected by the body's gravity.

How about before we sent any spacecraft to orbit the Moon?

The earth goes around the sun and the moon moves around the earth. So, the earth-moon system (also known as earth-moon tango) moves around the sun. The moon and earth revolve about their common centre of mass, known as barycentre, which is about 4,670 km (2902 miles) from the centre of the earth. The earth and the moon both revolve around this point every 27.3 days as the point revolves around the sun. The mass of the moon can be estimated by writing a mathematical expression describing momentum balance around the barycentre and knowing mass of earth and radius of earth and the moon.

Q18. What is the white circle that we see around the moon occasionally?

A18. The white circle is called a halo. The moon as well as the sun halos occur when there are high, thin clouds (mostly cirrus and cirrostratus clouds) across the sky. These clouds contain millions of tiny ice crystals, each of which is shaped like a tiny hexagon. The halo formation phenomena may be due to the refraction of light that passes through the ice crystals, or the reflection of light from crystal faces, or a combination of both effects. The light passing through these ice crystals is bent (refraction) at a 22-degree angle creating a halo 22 degrees in radius. Scientists call them 22-degree halos.

The prism effect of light passing through these six-sided ice crystals separates the light into its various colours, resulting in a halo tinged with very pale rainbow colours with red on the inside and blue on the outside. The phenomenon of a lunar halo is like a rainbow produced by sunlight and rain. Because moonlight isn't very bright, lunar halos are mostly colourless, but you might notice redder on the inside and bluer on the outside of the halo. These colours are more noticeable in halos around the sun.

In rare circumstances you could see a double halo, caused by less-than-perfect crystals. These crystals would not have the hexagon shape and would rather take on a different form (such as a triangle), causing a different sized halo.

A 22-degree halo around the moon
Credit: Stephen Balaban. Wikimedia Commons. December 17, 2013
https://commons.wikimedia.org/wiki/File:moon_22_degree_halo_atherton_ca.jpg
Figure in Public Domain

One weather folklore says a ring around the moon means rain soon. This is often proved true, as cirrus and cirrostratus clouds generally precede rain and storm systems.

NOTE: Cirrus clouds are delicate, feathery clouds that are made mostly of ice crystals while cirrostratus clouds are thin, white clouds that cover the whole sky like a veil.

❑

CHAPTER 2

Formation of the Moon

Q19. How did the moon form?

A19. The origin of the moon is still a question. We do not know where it has come from or how it was formed. Over the centuries, scientists have proposed several hypotheses/models about its formation. However, none of these are without any flaws. The debate is still on. There are four hypotheses/models. The first three are from the pre-Apollo era, while the fourth, after the Apollo programme.

1. The Double-Planet Hypothesis

This was proposed in 1873 by Edouard Albert Roche, a French astronomer and mathematician. This hypothesis proposes that the moon and the earth were formed individually from the nebula that formed the solar system. The moon was formed in the vicinity of the earth and in orbit around the earth.

2. The Fission Hypothesis

Sir George Howard Darwin, an English astronomer, proposed this hypothesis in 1879. It states that there was no moon in the beginning. The earth was slightly bigger and heavier than it is today and was spinning faster. The huge centrifugal force, due to its rapid rotation, ripped apart the early earth while it was still molten, and the moon was born. The present Pacific Ocean basin is the most popular site for the part of the earth from which the moon came.

3. The Capture Hypothesis

In 1909, an American astronomer T.J.J. See suggested that the moon was formed somewhere else in the solar system. Somehow, it wandered close to the earth and was captured by the earth's gravity.

4. The Giant Impact Model

In 1946, Reginald Aldworth Daly of Harvard University postulated that the moon was formed by an impact rather than centrifugal forces as suggested by Sir Darwin. In 1974/75, William K. Hartmann and Donald R. Davis proposed a more plausible model, generally known as the "Giant Impact Model (GIM)". Their model suggests that at the end of the planet formation period, several satellite-sized bodies had formed that could collide with the planets or be captured. A Mars size object, known as Theia (in Greek mythology, Theia was the mother of the moon goddess Selene), struck the earth just after the formation of the solar system. The impact ejected large volumes of heated material from the outer layers of both the objects earth and Theia. While some of this material escaped into space, the rest of it stayed in orbit, and consolidated to form the moon. Evidence from the moon rocks and simulations back

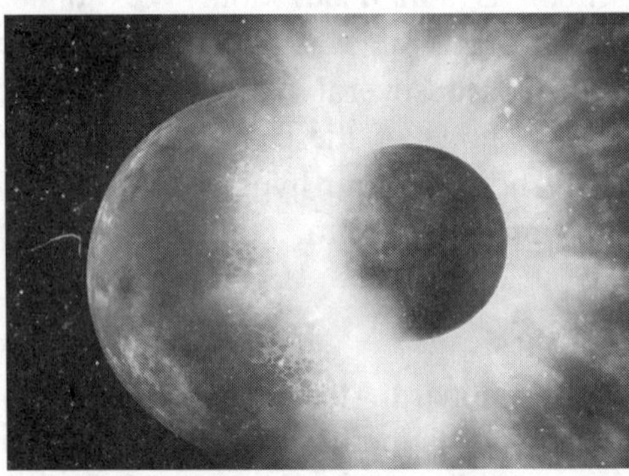

Artist's depiction of a collision between two planetary bodies. Such an impact between earth and a Mars-sized object likely formed the moon.
Credit: NASA/JPL-Caltech,
Created: August 26, 2009
https://www.nasa.gov/multimedia/imagegallery/image_feature_1454.html
https://commons.wikimedia.org/wiki/File:Artist%27s_concept_of_collision_at_HD_172555.jpg
Figure in Public Domain

up and support this idea. A similar approach was taken by Canadian astronomer Alastair G. W. Cameron and American astronomer William R. Ward, who suggested that the moon was formed by the tangential impact upon earth.

References

1. *Bobst, Kristen. 4 Theories About How the moon Formed. Treehugger.com Updated February 7, 2021* https://www.treehugger.com/theories-about-how-moon-formed-4863990
2. *Hartmann, William K and Davis, Donald R. Satellite-Sized Planetesimals and Lunar Origin. ICARUS 24,504-515(1975).* http://web-static-aws.seas.harvard.edu/climate/eli/Courses/EPS281r/Sources/Origin-of-the-moon/more/Hartmann-Davis-1975.pdf

Q20. What is the leading hypothesis/model of the moon's formation?

A20. An international meeting about the moon's origin was organized in Hawaii in 1984. The Giant Impact Model emerged as the leading one.

The Apollo missions of the USA and the USSR moon exploration programmes represent a crucial dataset. Using multiple measurements, they allowed us to test long-standing hypotheses of the moon formation. Thanks to hundreds of kilograms of samples that were brought back to earth.

Any model which explains the existence of the moon must explain the following:

1. The moon's low density (3.3 g/cc) suggests that the moon does not have a substantial iron core like the earth does.
2. Fewer volatile substances in the moon rocks implies extra baking of the lunar surface relative to that of the earth.

3. The relative abundance of oxygen isotopes on earth and on the moon suggests that the earth and moon were formed at the same distance from the sun.

(Isotope: All isotopes of an element have the same number of protons but different numbers of neutrons in their nuclei).

The Double-Planet hypothesis proposes that the moon was formed near earth. In that situation, it should possess a significant iron core like earth, but it does not. It is hard to imagine the two bodies growing together but somehow earth got all the stuff with iron in it.

The Fission hypothesis was thought possible since the moon's composition resembles that of the earth's mantle. A rapidly spinning earth could have thrown out the moon from its outer layers. When scientists calculated spin-speed required for such a thing to happen, they found it to be impossibly fast. That is why the Fission hypothesis was not considered favourably.

The Capture hypothesis also does not explain many things. The idea that the earth's gravity caught the moon might explain some of the compositional differences, but there are many similarities too. Also, why does the moon not have its own regular sized core, why is it smaller? Scientists think that such a fine tuning for capturing the moon and placing it in its present orbit is improbable. Something would have to slow it down by just the right amount at just the right time. If the moon had come in from elsewhere, it probably would have crashed into earth or escaped and fled into space.

Also, all the above three hypotheses do not have any explanation for the extra baking the lunar material has received.

Now let us look at the Giant Impact Model.

The Giant Impact Hypothesis is very good in explaining most of the moon's features.

If the material that formed the moon originated from the outer layers of the earth and Theia, rather than from their cores, an iron-poor moon would naturally result. The enormous impact energy associated with such an event would vaporize a substantial portion of the ejecta, accounting for the moon's lack of volatile materials. Regarding somewhat similarities in oxygen isotopes on the earth and the moon, some of the scientists, who looked at Apollo missions' lunar rocks, have suggested that the moon may be made up of about 40 per cent Theia. They have also proposed that Theia had a very similar isotopic composition to that of the earth. Such a composition of Theia would explain why no isotopic differences for oxygen, or any other isotope system had been detected to date.

Q21. What really happened during the moon's formation based on the Giant Impact Model?

A21. The giant impact between Theia and proto earth ejected small solid bodies which gathered and formed the moon. The rapid joining of the small bodies released heat melting the moon and creating an ocean of melted rock known as magma. Magma slowly cooled and solidified. As it cooled, dense, iron rich materials sank deep into the moon. Other materials also cooled and solidified forming the mantle (layer of rock beneath the crust). As the crust formed, asteroid bombarded it heavily shattering and churning it and creating craters. The largest impacts may have stripped off the entire crust. One such collision created the South-Atkin Basin, one of the largest known impact craters in the solar system. About 4 billion to 3 billion years ago, melting occurred in the mantle, probably caused by radioactive elements deep in the moon's interior. The resulting magma erupted as dark, iron-rich lava, partly flooding the heavily cratered surface. The lava cooled and solidified into rocks known as basalts. Small eruptions may have continued until as recently as 1 billion years ago.

This is illustrated in the following figure

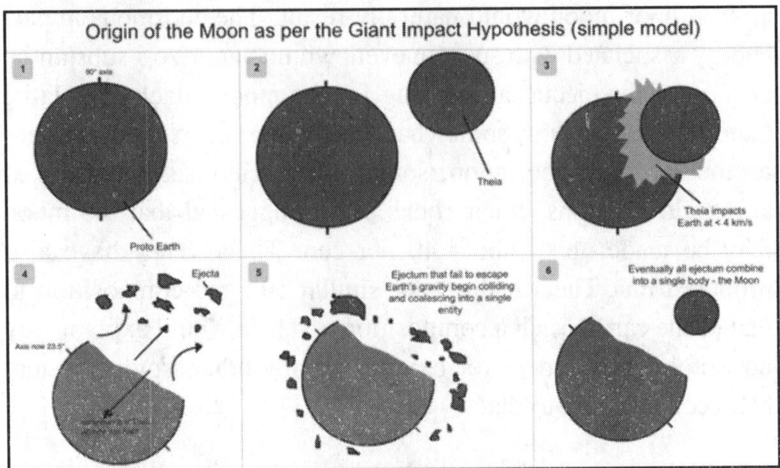

Simplistic Representation of the Giant-Impact Hypothesis.
Credits: Jatan Mehta, July 19, 2018 (modification of the original image by Bedrock Person)
https://www.planetary.org/articles/apollo-moon-origin

Reference

Mehta, Jatan. How the Apollo missions transformed our understanding of the Moon's origin. The Planetary Society. July 19, 2018
https://www.planetary.org/articles/apollo-moon-origin

Q22. Is the Giant Impact Model of the moon's formation the final answer?

A22. No. Although the Giant Impact Model does a better job than the other theories, there are issues. Even William Hartmann, who proposed the Giant Impact Model in the 1970s along with Donald R. Davis, has written in a blog that the Giant Impact Hypothesis is not yet settled. Scientists are in a constant endeavour finding new models.

Some of the issues in the Giant Impact Model are:

1. Scientists have a hard time understanding a scenario where Theia hits earth with just the right amount of force. The collision had to be strong enough (a) to throw a huge amount of material up into orbit around the earth, but (b) not so strong as to destroy the earth.

3. Scientists also question why the debris formed into a single moon rather than a few smaller moons.

4. Most of the bodies of the solar system have unique chemical makeups. The earth and Theia should also be like this. However, the Apollo and Luna lunar samples show that the moon and earth have very similar compositions. Apart from their lack of iron and extreme lack of water, moon rocks match earth's isotopic ratios for titanium, calcium, silicon, oxygen, and tungsten (isotopes of an element have same numbers of protons but differing numbers of neutrons from one another). What happened to the contribution of Theia? Additionally, only in rare cases, 1 per cent or 2 per cent of the time, do simulations yield a moon with an earthlike composition. There is also a problem of fine-tuning the impact to yield the angular momentum of the current earth-moon system.

   In defence, some scientists have suggested that Theia could have been made of ice. Or, if Theia formed at around the same distance from the sun as earth, it would contain mostly the same material. A potential explanation comes in the form of a rare group of meteorites, called enstatite chondrites, which are also nearly identical to earth. These meteorites may have been the building blocks of earth and Theia as well.

5. Scientists, based on their simulations and chemical analysis of the moon and earth rocks, have differing opinions about Theia's contribution to moon formation. While a few suggest that most of the material (60 to 80 percent) which formed the moon has come from Theia rather than from earth, others

(Herwartz and others) say that the moon may be made up of about 40 percent Theia. Some do not agree with either of these findings.

6. There is also a debate about how the impact happened. Some computer simulations suggest that Theia struck earth at an estimated 45-degree angle. Originally, the hypothesis supposed that Theia had struck earth with a glancing blow. Another publication suggests that the impact was indeed a head-on collision and that Theia's remains can be found in both the earth and the moon.
7. A few years ago, some of the simulations seemed to show that Theia had originated in the distant solar system, complicating the hypothesis further.

So, the debate is on. It is therefore evident that further refinement in existing moon formation models will continue, and new models will emerge to get the final theory of the moon formation. But currently, Giant Impact Theory is the dominant hypothesis about the moon formation. Further progress in the field will be achieved by the synergy of astrophysics and planetary sciences jointly with computer modelling.

References

1. *Hartmann, William K. What happened after the giant impact that created the Moon? Astronomy. Home/ Magazine/Ask Astro/Ask Astro. Published: Wednesday, March 24, 2021*
https://astronomy.com/magazine/ask-astro/2021/03/ask-astro-what-happened-after-the-giant-impact-that-created-the-moon
2. *Herwartz, Daniel and Pack, Andreas and Friedrichs, Bjarne and Bischoff, Addi. (2014). Identification of the giant impact Theia in lunar rocks. Science. 344. 1146-1150. 10.1126/science.1251117.*

Q23. What are the other ideas scientists are now working on to explain the moon's formation?

A23. Scientists are working on other models. Three such hypotheses are:

1. Multiple Impact Hypothesis
2. First the moon, Then earth Hypothesis
3. One-Two Punch One-Two Punch Hypothesis

1. Multiple Impact Hypothesis for the Moon's Formation

The commonly accepted Giant Impact Model suggests that the one big collision threw up debris that eventually coalesced and formed the moon. This theory is supported by chemical analyses of moon rocks showing them to be pretty much identical to earth rocks. But in all the moon samples, there is no trace of Theia, that hit the proto earth. If a large body did hit the earth, there should be traces of that body among the debris that formed the moon. In 2017, Israeli scientists, Rufu, Aharonson, and Perets suggested that the ancient earth once hosted a series of moonlets, each one formed from a different collision with the proto earth. Such moonlets were later ejected or collided with earth or with each other. A long series of such moon-moon collisions formed a bigger moon - the moon we see today. This model is consistent with the current understanding of the formation of the earth. In its last stages of growth, earth experienced many giant impacts with other bodies. Each of these impacts contributed more material to the proto earth, until it reached its current size. This Multiple Impact Hypothesis helps to explain the chemical composition similarities between earth and the moon. It has been reported that this idea was first proposed in 1989, it could not be tested then due to lack of computer power.

In another publication, Robert Citron and others have also argued that a single giant impact is less probable. They have suggested that in the multiple impacts model, a pre-existing moonlet tidally evolves outward till a subsequent impact forms a new moonlet, at which point both moonlets will tidally evolve until a merger or system disruption. Formation of the moon from the mergers of several moonlets could therefore be a natural byproduct of the earth's growth through multiple impacts. They have explained their model (computer simulations) in the following figure.

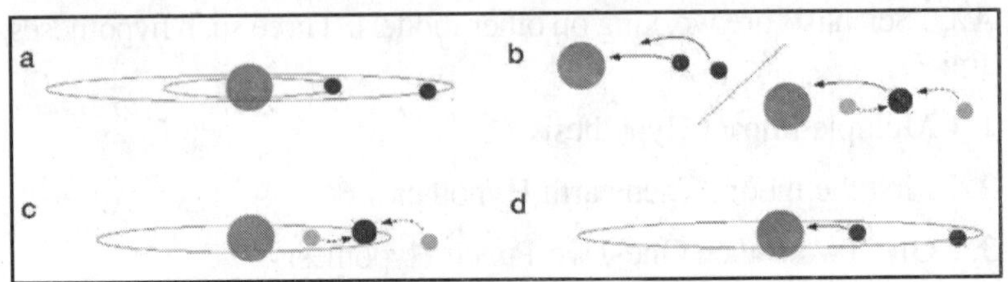

Potential outcomes from computer simulations of Robert Citron and others (June 1, 2018)

The system begins with: (a) an outer moonlet from a previous impact/merger that has migrated to a distant orbit, and an inner moonlet produced by a recent impact. The system evolves until either (b) both moonlets in-fall individually or after a merger, impacting the proto-earth (c) the moonlets merge into a larger moonlet that remains stable and continues to tidally evolve outward, or (d) the inner moonlet in-falls while the outer moonlet remains stable.

The Figure is from the paper:

https://iopscience.iop.org/article/10.3847/1538-4357/aaca2d

Credits: Reproduced with the permission of Robert I. Citron.

2. First the Moon, Then Earth Hypothesis

This model is also known as "The Origin of the moon Within a Terrestrial Synestia". The idea was put forward by Simon Lock and Sarah Stewart in 2018.

Astronomers have proposed a whole new type of planetary object called a synestia ("syn", means "together" and "Estia," Greek goddess of architecture and structures). Synestia is a huge, spinning, doughnut-shaped mass of hot, vaporized rock, formed early in a planet's life, when large (planet-sized) objects smash into each other, also applicable to earth.

According to Lock, Stewart, and others, the moon was formed prior to the earth. A giant impact drove the earth into a synestia. The moon formed as the synestia shrank and cooled. The earth subsequently emerged about 1,000 years after the moon. The moon orbited within the synestia long enough to chemically equilibrate with the vaporized earth. Over time, the whole structure shrank, and the moon emerged from the vapor. Eventually, the whole synestia condensed and what's left is a ball of spinning, liquid rock that eventually formed the earth. This model matches the pattern of the moon's composition. The moon inherited its composition from the earth, but because it formed at high temperatures it lost the easily vaporized elements, explaining the moon's distinct composition.

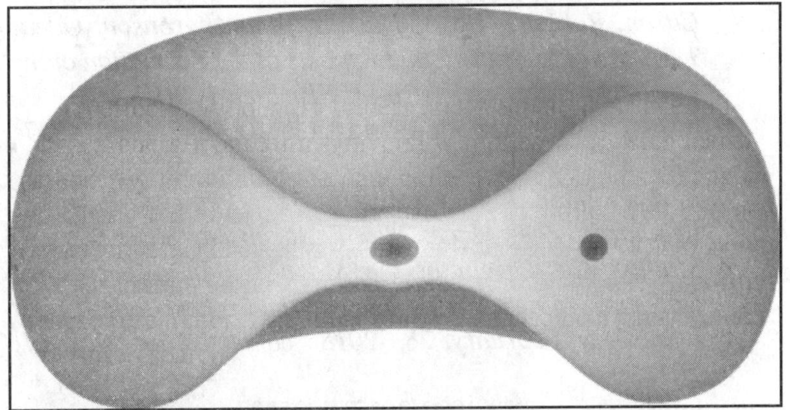

The moon Forming in a Synestia

The dark smudge in the centre of this illustration is approximately earth-sized, showing the immense size of a synestia. The interior consists of rock vapour.

Credits: Reproduced with the permission of Simon J. Lock. Date: 2018

3. One-Two Punch Hypothesis

In a blog, ASA Stahl has discussed a recent study (October 2021) by Asphaug and colleagues according to which two impacts, not just one, may have formed the moon. Theia most likely hit earth twice, the first time only glanced off earth and some hundreds of thousands of years later, it came back to deliver the final blow. Initially, Theia could have been going much faster and the first impact would have slowed it down to the perfect speed for the second one. A hit-and-run return would enable earth's and Theia's materials to mix more than in a single impact, ultimately forming a moon chemically more similar to earth. In their research publication, the proposers of this hypothesis have indicated that more advanced simulations would yield even better results.

References

1. *Rufu, R., Aharonson, O., and Perets, H. A Multiple-Impact Origin for the Moon. Nature Geosci 10, 89–94. January 9, 2017.* https://doi.org/10.1038/ngeo2866
2. *Citron, Robert I, Perets Hagai B, and Aharonson, Oded. The Role of Multiple Giant impacts in the Formation of the Earth-Moon System. astro-ph.EP. June 1, 2018* https://iopscience.iop.org/article/10.3847/1538-4357/aaca2d
3. *Lock , Simon J. Stewart, Sarah T. Petaev, Michail I. Leinhardt, Zoë. Mace, Mia T. Jacobsen, Stein B. Cuk, Matija. The Origin of the Moon Within a Terrestrial Synestia. Journal of Geophysical Research. First published: February 28, 2018. corrected on May 21, 2018.* https://doi.org/10.1002/2017JE005333
4. *Stahl, ASA. The Moon may have formed in a one-two punch. Sky & Telescope. October 14, 2021* https://skyandtelescope.org/astronomy-news/two-impacts-not-just-one-may-have-formed-the-moon/
5. *Asphaug, Erik. et al. Collision Chains Among the Terrestrial Planets. III. Formation of the Moon. The Planetary Science Journal, Volume 2, Number 5 (Planet. Sci. J. 2 200). September 23, 2021* https://iopscience.iop.org/article/10.3847/PSJ/ac19b2

Q24: According to the Giant Impact Model of the moon's formation, the moon is formed from a collision of earth and Theia. What happened to the rest of Theia after the collision?

A24. When the Giant Impact Hypothesis was proposed by Hartmann and Davis (1970), not much information was provided about the impactor Theia. In the years since, many researchers have modelled what the impact may have looked like and what might have happened to Theia.

Earlier models suggested that most of Theia's material ended up as part of the moon. Any iron core that Theia may have had was consumed by earth's own core.

Based on new isotopic evidence and modelling, a recent study (2021) led by Qian Yuan suggests that the remnant of Theia is still inside earth, probably located in two continent-sized layers of rock beneath West Africa and the Pacific Ocean. Scientists have observed that these two rock layers are denser and chemically different from the surrounding mantle rock as they slow down seismic waves from earthquakes. Seismologists call these rocks Large Low-Shear Velocity Provinces (LLSVPs). They are the largest thing in the earth's mantle, up to 1000 kilometres tall and several times that wide. As mentioned in Voosen's news item published on the Science news website "Yuan's model suggests that after the collision, Theia's core would have quickly merged with earth's. He also probed the fate of Theia's mantle and found out that mantle rocks 1.5 per cent to 3.5 per cent denser than earth's would survive and end up as piles near the core rather than mixing in". There are limitations in this model, and it is being discussed among various scientists.

References

1. *Hartmann, William K. Ask Astro: What happened after the giant impact that created the moon? Astronomy. March 24, 2021.*
   https://astronomy.com/magazine/ask-astro/2021/03/ask-astro-what-happened-after-the-giant-impact-that-created-the-moon
2. *Yuan, Q. Li, M. M. Desch, S. J. and Ko, B. Giant Impact Origin for the Large Low Shear Velocity Provinces.*

Arizona State University, School of earth and Space Exploration, PO Box 876004, Tempe, AZ85287-6004, USA. 52nd Lunar and Planetary Science Conference 2021 (LPI Contrib. No. 2548)
https://www.hou.usra.edu/meetings/lpsc2021/pdf/1980.pdf

3. *Voosen, Paul. Remains of impact that created the moon may lie deep within earth. Science. Mar. 23, 2021*
https://www.sciencemag.org/news/2021/03/remains-impact-created-moon-may-lie-deep-within-earth

Q 25. Are there any significant contributions made by Theia's collision to proto earth that made the moon?

A25. Some of the contributions are given below:

(1) Without the impact that formed the Moon, we might not have life on earth.

The earth was not formed containing the necessary chemicals for life to begin. One theory, called the "late veneer theory", suggests that the volatile chemicals needed for life were brought to the earth by meteorites. But studies showed that the earth's volatiles are not in proportions that are expected from addition of chondrite-like meteorites. Based on these observations, a new study by Grewal, Dasgupta and others of the Rice University, USA (as reported by Jade Boyd of Rice University and Evan Gough of Universe Today) has challenged this theory. Using laboratory experiments and computer simulations, the Rice University team has argued that the impacting object, Theia had a sulfur-rich core, while its mantle and crust contained volatiles. When it collided with the earth, it injected the chemicals necessary for life, like nitrogen, carbon, hydrogen, and sulfur, into the earth's silicate reservoirs (mantle + crust). The collision also ejected massive amounts of material into space, which coalesced to form the moon.

> **Note**: Chondrites constitute primitive meteorites, which represent chemical properties of the solar nebula, the cloud of gas and dust from which the solar system formed.

(2) The collision that created the Moon might have also brought water to the early earth.

Regarding the presence of water on the earth, standing theory says that earth gathered its water from comets and asteroids. But a new recent study by Budde, Burkhardt, and others from the University of Munster, Germany, based on molybdenum isotopes, presents evidence that supports a different source for earth's water: Theia itself. Their explanation is that Theia, which collided with earth 4.5 billion years ago forming the moon, may have originated in the outer Solar System rather than in the inner solar system, bringing water and carbon-based materials with it. According to the scientists, the collision provided the entire amount of water on earth.

References

1. *Gough, Evan. Without the Impact that Formed the moon, We Might Not Have Life on earth. Universe Today. January 29, 2019* https://www.universetoday.com/141342/without-the-impact-that-formed-the-moon-we-might-not-have-life-on-earth/
2. *Boyd, Jade. Planetary Collision that Formed the Moon Made Life Possible on Earth. Posted in Current News, Featured Stories, Rice University-News and Media Relations (Office of Public Relations). January 23, 2019.* https://news2.rice.edu/2019/01/23/planetary-collision-that-formed-the-moon-made-life-possible-on-earth-2/
3. *Gough, Evan., The Collision that Created the Moon Might Have Also Brought Water to the Early Earth. Universe Today. May 23, 2019* https://www.universetoday.com/142290/the-collision-that-created-the-moon-might-have-also-brought-water-to-the-early-earth/
4. *Budde, Gerrit. Burkhardt, Christoph. and Kleine Thorsten. Molybdenum isotopic evidence for the late accretion of*

outer Solar System material to Earth. Nature Astronomy (IF 11.518) Pub Date : 20-May-2019
https://www.nature.com/articles/s41550-019-0779-y

Q26. How did the early civilizations explain the moon's birth?

A26. Before scientists started theorizing the moon's formation, the early civilizations, all along, strived to explain the birth of the moon in terms of myths and legends. There are numerous stories world over. Three of them (Hindu mythology, Philippines, and Africa) are given below:

1. **In one of the episodes in Hindu mythology, the moon, known as Chandra, was created out of the cosmic ocean of milk. When the gods and demons were busy churning the ocean, several objects yielded out of the process. Following many valuables, the moon finally floated to the surface and kept on rising in the air. The gods decided that the moon should find a place amongst the stars in the sky. The moon has been circling the heavens since then.**
2. **According to a Philippine folk tale, in earlier times the sky looked like a coral rock and was close to the ground. One day, a spinster went out to pound rice. She hung her neck-beads and comb on the sky before she started pounding rice. The sky was so low that each time she raised her pestle into the air it struck it. After some time, she raised the pestle so high that it struck the sky extremely hard. The sky started rising. It went up so far that she lost her ornaments. The comb became the moon and the beads scattered and became the stars.**
3. **In an African folklore, the great creator Mbombo (also called Bumba) created the universe from nothing. The story affirms that in the beginning, Mbombo was alone, darkness and primordial water covered all the earth. Once Mbombo felt an intense pain in his stomach and vomited the sun, the moon, and stars. The heat and light from the sun evaporated the water covering the**

earth, creating clouds, and after time, the dry hills emerged from the water. Then Mbombo vomited once more, bringing forth nine animals: a leopard, an eagle, a crocodile, a fish, a tortoise, a black leopard-like animal a white heron, a scarab, and a goat. Mbombo also vomited many men, one of them was called Loko Yima. These nine animals went on to create all the world's creatures. Three of Mbombo's sons created the rest of the world: white ants, the first plant (which in turn gave rise to all trees, grasses, and flowers), and the last bird. Once the creation was complete and peaceful, Mbombo delivered it to mankind and retreated into the heavens, leaving Loko Yima to serve as "god upon the earth". The woman of the waters, Nchienge, lived in the East, and her son, Woto, became the first king of the Kuba (in Africa).

Note: Around 2500 years ago, the Greek philosophers announced that the way to understand all-natural processes was through reason and not myths or legends. However, truly modern scientific study began in 1609, when Galileo Galilei, an Italian astronomer and philosopher, looked at the moon through his telescope. He saw vast plains, mountains, and valleys similar to those on the earth. Over the centuries that followed, scientists proposed a variety of theories about the moon's origin. And the work is still on.

Reference

Mabel, Cook Cole. How the Moon and the Stars Came to Be. Philippine Folk Tales. A. C. McClurg and Company, Chicago. 1916, p. 124.

Q27. How old is the moon?

A27. We do not have a clear answer. It is a subject of a debate within the scientific community. Some say that it was formed roughly 100 million years after our solar system, while others favour a date somewhere between 150 and 200 million years after the solar system's birth. This means that the moon's age is between 4.35 and 4.47 billion years.

However, new research by Thiemens, Sprung and others from the University of Cologne's Institute of Geology and Mineralogy suggests that the moon is 100 million years older than what earlier thought. Their research is based on chemical analyses of Apollo lunar samples which shows that the moon formed only 50 million years after the solar system. Therefore, according to their findings, after the formation of the solar system, 4.56 billion years ago, the moon formed approximately 4.51 billion years ago.

Another group of scientists from Planetary geophysicists at the German Aerospace Center (Deutsches Zentrum für Luft- und Raumfahrt; DLR in short), led by Maxime Maurice, together with researchers at the University of Münster, have used a new numerical model to reconstruct the time at which the event occurred—4.425 billion years ago.

Scientists still believe that determining the age of the moon is an overly complicated problem and we still do not have a clear answer to such an obvious question.

References

1. *Thiemens, M.M., Sprung, P., Fonseca, R.O.C. et al. Early Moon formation inferred from hafnium–tungsten systematics. Nat. Geosci. 12, 696–700 (2019).* https://doi.org/10.1038/s41561-019-0398-3
2. *Gough, Evan. The moon is Older Than Scientists Thought. Universe Today. July 31, 2019.* https://www.universetoday.com/143025/the-moon-is-older-than-scientists-thought/
3. *DLR. Researchers find younger age for Earth's moon. PHYS-ORG, July 13, 2020.* https://phys.org/news/2020-07-younger-age-earth-moon.html

Q28. How do scientists estimate the moon's age?

A28. Scientists estimate the moon's age by using radioactive decay of elements like uranium and zirconium. This is known as radiometric dating. All rocks and minerals contain long-lived radioactive elements that were incorporated into the moon, Earth, other planets, meteorites, etc. when the solar system was formed. By taking an element with a known rate of decay, and knowing its concentration in moon rocks, researchers can back-calculate a time for when the material formed. Samples for dating are selected carefully to avoid those that are altered, contaminated, or disturbed by later heating or chemical events. ❑

CHAPTER 3

Internal Structure of the Moon

Q29. What is the internal structure and composition of the moon?

A29. The moon is composed of layers with different compositions. The heaviest materials have sunk into the moon's centre, and the lightest materials have risen to the outermost layer. Just like our earth, the moon has a core, mantle, and crust. Although, the deep structure of the moon is still debated due to the absence of comprehensive seismic data, scientists have explored lunar interior models by complementing Apollo seismic data with Gravity Recovery and Interior Laboratory (GRAIL) and Lunar Laser Ranging (LLR). They have also used seismic, rotational, and gravity measurement studies to gain insights into the different layers within the moon. The general pictures is like this.

At the centre is the dense, metallic core, largely composed of iron and some nickel. The inner core is a solid mass surrounded by a fluid outer core. The moon's core is small (less than 20 per cent of the Moons diameter) as opposed to other terrestrial worlds (like the earth) with cores measuring closer to 50 per cent of their diameters. Above the core are the mantle and crust. The mantle is far more extensive than the crust. Interestingly, the crust of the moon seems to be thinner on the side of the moon facing the earth (near side) and thicker on the side facing away (far side). Researchers are still working to determine why this might be. Data from GRAIL mission suggest that the crust of the moon is thinner than we had thought ascertaining the thickness of its crust to be 34-43 km (21-27mi).

In the beginning the moon was largely composed of a great ocean of magma. As the magma ocean began to cool, crystals began to form within the magma. Crystals of denser mantle minerals, such as olivine and pyroxene (which are made up of magnesium, iron, silicon, and oxygen atoms) sank down to the bottom of the ocean. Lighter minerals, notably anorthositic plagioclase feldspar (which are made up of oxygen, silicon, magnesium, iron, calcium, and aluminium, with small amounts of titanium, uranium, thorium, potassium, and hydrogen) crystallised and floated to the surface to form the moon's crust.

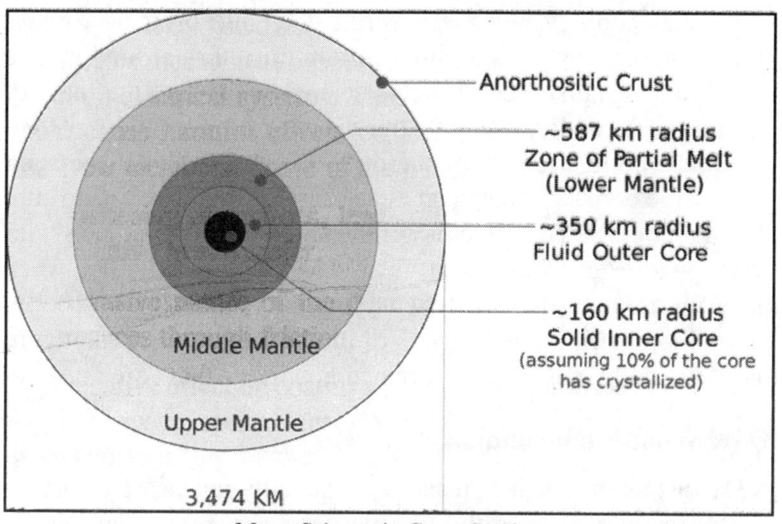

Moon Schematic Cross Section
Credits: Author: Bryan Derksen at English Wikipedia
Date 15 April 15, 2007 (original upload date)
https://commons.wikimedia.org/wiki/
File:moon_Schematic_Cross_Section.svg
Figure in public domain

Q30. How do we know about the internal structure of the moon?

A30. Seismographs make studying any celestial body's interior possible. These instruments record waves that travel through the interior of a body which help to determine the structure and composition of different layers of the body. Just as geoscientists study earthquakes to learn about the structure of earth, seismic waves of "moonquakes" (seismic events on the moon) can be analysed to probe the lunar interior. NASA deployed the first seismographs on the moon as part of the Apollo Mission in 1969. These seismographs collected data and enabled researchers to determine the moon's structure consisted of a thin crust, a mantle, and a core.

In absence of seismic data, the interior of a planet or moon (celestial body) can be determined from spacecraft navigation data, that is when a spacecraft passes by or goes into orbit around a celestial body. When a spacecraft goes into orbit of a body, the body's gravity pulls it into a certain trajectory around it. The trajectory of the spacecraft helps scientists estimate the mass of the body. Secondly, the way the mass of a body is distributed inside the body affects how the body spins in space. If the body has a large core, it will spin at a faster speed than if it has no core or smaller core. The effect is somewhat like that of an ice skater who spins faster when her arms are close to the body than when the arms are spread apart. By studying the rate of spin of a body, as well as determining the mass of the body, scientists can figure out if the body must have a core and how large that core must be.

Q31. What is a moonquake?

A31. Just as the earth's surface occasionally quakes, the surface of the moon also goes through periodic shakes and quakes. A moonquake is the moon equivalent of an earthquake, a kind of vibration on the surface of the moon. They were first discovered by the Apollo astronauts. Astronauts placed five seismometers on the moon's surface during the Apollo 11, 12, 14, 15 and 16 missions. The Apollo 11 seismometer operated only for three weeks, but the four remaining instruments recorded 28 shallow moonquakes from 1969 to September 1977. Apollo 11 astronauts were the first ones who observed a quake there. The seismometer recorded the first moonquake minutes after the seismometer had settled down and transmitted the first ever seismic data from the moon back to earth.

Q32. How does a moonquake differ from an earthquake?

A32. Moonquakes and earthquakes are different in magnitude, lasting times, and sources. Some of the important differences are:

1. **Causes:** Earthquakes are caused due to tectonic plates collision, volcanoes, and man-made explosion etc. The moon does not have tectonic plates. Moonquakes are due to tidal forces, moon's interior cooling, meteorite impact, and thermal impact etc. Most of the moonquakes are concentrated at great depth—between 600 km and 1000 km, which is deeper than earthquakes.
2. **Quake sizes:** Intensity of the moonquake is smaller than the earthquake, but the occurrence time (up to an hour) is larger than earthquake due to fewer attenuating factors to dampen seismic vibrations. The total seismic energy release within the moon appears to be about 80 times less than that in the earth. The largest earthquake ever recorded was a magnitude 9.5 on the Richter scale on May 22, 1960, in Valdivia, Chile. Between 1972 and 1977, the Apollo seismic network saw twenty-eight moonquakes; a few registered up to 5.5 on the Richter scale.
3. **Quake duration and destructive effects:** The earthquake has an enormous destructive effect on earth causing great harm to life and property. Quakes of the magnitude of 4.5 can cause damage to a rigid structure on the earth. Moonquakes are also longer than earthquakes.
4. **Prediction:** While significant works are being done on earthquake forecasting, a reliable and skilful earthquake prediction is not yet possible, most moonquakes (except

surface moonquake) are predictable as they are regular in time occurrence and related to tidal stresses and position of the moon and earth.

Q33. What are the different types of moonquakes and what are their causes?

A33. There are at least four different kinds of moonquakes as mentioned below. The actual causes for moonquakes are not yet known. Each type of moonquake have a distinct cause rooted in the moon's structure and its position in our solar system.

1. Deep moonquakes about 700 km below the surface, probably caused by the tides which are due to stretching and relaxing of the gravitational pull between the earth and the moon.
2. Meteor impacts vibrations caused when meteors crash into the surface of the moon.
3. Thermal quakes which are caused by the expansion of the cold crust when first illuminated by the morning sun after two weeks of deep-freeze lunar night
4. Shallow moonquakes, only 20 or 30 kilometres below the surface, likely caused when the moon's crust slips and cracks due to the gradual shrinking of the moon as it cools. This shrinking is sometimes referred as "raisining" to express similarity with formation of resins by drying up grapes.

The first three are generally mild and harmless. Shallow moonquakes on the other hand are problematic.

Q34. Does the "Moon shrinking" cause moonquakes?

A34. The moon is shrinking as its interior cools, getting more than about 150 feet (50 metres) skinnier over the last several hundred million years. It contracted and wrinkled like a raisin as its interior cooled. Unlike the flexible skin on a grape, however, the moon's crust is brittle, causing it to break as the interior shrinks. This breakage results in thrust faults, where one section of crust is pushed up over an adjacent section. This has left behind cliffs which are known as thrust faults on the moon's surface.

Based on the seismic data from instruments placed on the moon's surface by Apollo missions and Images from NASA's Lunar Reconnaissance Orbiter (LRO), scientists suggest that these faults are still active and likely producing moonquakes today as the moon continues to gradually cool and shrink. The team has produced "shake maps" that predict where the strongest shaking should occur, given the size of the thrust faults. This information will be particularly useful for future trips to the moon. The researchers also found that most of the quakes happened when the moon was at or near its apogee, the point in the moon's orbit when it is farthest from earth. This is where additional tidal stress from earth's gravity causes a peak in the total stress on the moon's crust, making slippage along the thrust faults more likely.

Q35. Are Moonquakes dangerous to the astronauts and equipment?

A35. Shallow Moonquakes are the most powerful and the most worrisome for researchers and those eager to colonize the moon. These could do some real damage. Between 1972 and 1977, the Apollo seismic network saw twenty-eight of them - a few registered up to 5.5 on the Richter scale. A magnitude 5 quake on earth is energetic enough to move heavy furniture and crack plaster. Furthermore, shallow moonquakes lasted a remarkably long time. Once they got going, they all continued more than 10 minutes. For a lunar habitat, that persistence could be more significant than a moonquake's magnitude.

Q36. Why do Moonquakes last for a longer time compared to earthquakes?

A36. Quakes on earth tend to last between 10 and 30 seconds, and some may last up to two minutes. The longest ever recorded was nearly ten minutes, but that is extremely rare. Moonquakes on average last far longer than earthquakes. The reason has to do with water weakening stone on earth. When the tectonic plates rumble and an earthquake occurs, the huge amount of energy spreads out through the mineral-rich crust of the earth, which has largely been infiltrated by water in the stone. This makes the material slightly more compressible, and able to absorb energy and seismic waves, diminishing the power and slowing the shaking. On the moon, things are quite different. The moon is dry, cool, and mostly rigid. So, moonquakes set it in vibrating mode. It just keeps going endlessly. It is not uncommon for a moonquake to last for ten minutes or more, and for smaller trembling to continue for hours afterwards.

Q37. Does the moon have volcanic activity now?

A37. The moon has been volcanically active throughout much of its history, with the first volcanic eruptions having occurred about 4.2 billion years ago. Volcanism was most intense between 3.8 and 3 billion years ago, during which time much of the lunar lava plains were created. This activity was originally thought to have gradually slowed down and died about 1 billion years ago. It is true that the moon is not very active nowadays and has not been very active for a long time. But scientists have found evidence for dozens of burps of volcanic activity, all within the past 100 million years and they think that future eruptions are likely—although probably not within a human lifetime. Scientists at Arizona State University, Tempe, USA have stated "The big story is that the moon is warmer than we thought,". In 2006 they found evidence that a small crater, called Ina, was indeed a volcanic vent that was active as recently as 10 million years ago. They have found evidence for 70 young volcanic structures on the moon's near side. The team says that the flows must be relatively new. Therefore, the moon has not cooled as quickly as scientists once thought and they suggest that the moon is likely still experiencing quakes to this day.

Matt Williams, in his recent blog published in *Universe Today*, has mentioned that according to a recent study (October 2021) of the samples returned to earth by China's Chang'e 5 spacecraft, led by the Chinese Academy of Geological Sciences (CGAS), an international team of scientists concluded that lunar volcanism was still active until about 2 billion years ago.

References

1. *Billah, Moushtakim. Islam, Md. Mofizul. Ali, Rubieyat Bin (Bangladesh University of Engineering & Technology, Daca, Bangaladesh). Earthquake Vs. Moonquake: A Review. World Scientific News. WSN 100 (2018) 1-15. EISSN 2392-2192*
   http://www.worldscientificnews.com/wp-content/uploads/2018/04/WSN-100-2018-1-15.pdf
2. *Nakamura, Y. Shallow Moonquakes: How They Compare with Earthquakes, Lunar and Planetary Science Conference, 11th, Houston, TX, March 17-21, 1980, Proceedings. Volume 3. (A82-22351 09-91) New York, Pergamon Press, 1980, p.p 1847-1853. 1980.*
   https://adsabs.harvard.edu/full/1980LPSC...11.1847N
3. *Moonquakes. science-news/science-at-nasa. March 15, 2006: Author: Trudy E. Bell | Production Editor: Dr. Tony Phillips | Credit: Science@NASA*
   https://science.nasa.gov/science-news/science-at-nasa/2006/15mar_moonquakes
4. *Hand, Eric. Recent volcanic eruptions on the moon. Space. October 12, 2014.*
   https://www.science.org/content/article/recent-volcanic-eruptions-moon
5. *Sutter, Paul M. The Most Recent Volcanic Activity on the Moon? Just 100 Million Years Ago. Universe Today. March 9, 2021.*
   https://www.universetoday.com/150429/the-most-recent-volcanic-activity-on-the-moon-just-100-million-years-ago/
6. *Watters, T.R. Weber, R.C. Collins, G.C. et al. Shallow seismic activity and young thrust faults on the moon.*

Nature Geoscience. 12, 411–417 (2019). May 13, 2019. https://doi.org/10.1038/s41561-019-0362-2

7. Williams, Matt. Volcanism on the Moon Ended About 2 Billion Years Ago. *Universe Today*. October 17, 2021. https://www.universetoday.com/152906/volcanism-on-the-moon-ended-about-2-billion-years-ago/

❑

CHAPTER 4

Lunar Surface and the Moon Mapping

Q38. What are the three main surface features on the moon?

A38. The surface of the moon is quite different from the earth. It is like a desert characterized by many visible features. The three main features are: dark and light areas, impact craters, and fine dust.

The dark features of the moon are called maria. Early scientists thought that the dark areas of the moon were oceans, and so erroneously named such features mare, which is Latin for «seas» (plural maria). Maria are the largest topographic features on the moon and can be seen from earth with the unaided eye. The light areas are known as the highlands on the moon. These light and dark areas represent rocks of different composition and ages and that is why these areas appear dark grey and white.

With too sparse an atmosphere to block impacts, a steady rain of asteroids, meteoroids and comets strikes the surface of the moon, leaving numerous craters behind.

Over billions of years, these impacts have grounded up the surface of the moon into fragments ranging from huge boulders to powder. Nearly the entire moon is covered by a rubble pile of charcoal-grey, powdery soft dust and rocky debris called the lunar regolith.

The image below illustrates some of the moon's features which can be seen from earth.

Moon Features You Can See from Earth
Credit: NASA/GSFC/Arizona State University/The Planetary Society
Figure is free for printing and distribution with attribution
Reproduced with the permission
https://www.planetary.org/space-images/moon-features-you-can-see-from-earth-square

Q39. What are other important surface features on the moon, other than the three main features (dark and light areas, craters, and moon dust)?

A39. There are several other surface features on the moon. The International Astronomical Union (IAU) segregates these features into 18 categories based on the current system of lunar nomenclature. Some of the features are: valleys, Terrae (brighter hilly, mountainous areas that are higher on the moon's surface than dark areas), Rimae (Lunar rilles which are long thin tunnels or trenches), dorsa (a wrinkle-ridge system commonly found on lunar dark areas), areas with higher albedo (Geographic areas distinguished by amount of reflected light), Promontoria (headlands or capes extending from a larger upland on the moon), Rupes (escarpments in the surface),

Dark Areas on the Moon

Q40. What are dark areas on the moon?

A40. The moon's dark areas are known as maria. These are lava-filled basins which were dubbed maria, Latin for "seas", by early astronomers who mistook them for actual seas. Of course, we now know that there are no seas. But the names stayed. For instance, the first Apollo landing occurred in Mare Tranquilitatis (the Sea of Tranquillity). Maria basins are less reflective because of their iron-rich composition, and hence appear dark to the naked eye. Areas of mare material occupy about 16 per cent of the moon's total surface, mostly on the near side of the moon, the side visible from earth.

Q41. When and how were the maria basins formed?

A41. The maria basins were formed beginning about 3.9 billion years ago. That was a period of intense bombardment by asteroid-sized bodies well after the lunar crust had cooled and solidified enough to retain large impact scars. Then, over a period lasting until perhaps three billion years ago, a long sequence of volcanic events flooded the giant basins and surrounding low-lying areas with magma originating hundreds of kilometres within the interior.

Q42. How do scientists determine the age of the moon maria?

A42. Ages of the maria are being determined by two methods:

(1). Absolute ages are given by radiometric dating techniques of the rocks from maria.

(2). Relative ages can be established by counting craters in mare surfaces. Crater counting is a method for estimating the age of a planet's surface. The method is based upon the assumption that when a piece of surface is new, then it has no impact craters and impact craters accumulate after that at a rate that is assumed known. Consequently, counting how many craters of various sizes there are in a given area allows determining how long they have accumulated and, consequently, how long ago the surface has formed. The method has been calibrated using the ages obtained by radiometric dating of samples returned from the moon by the Luna and Apollo missions. Besides moon maria surfaces, this technique has been used to estimate the age of areas on Mars and other planets that were covered by lava flows and how long-ago areas on the icy moons of Jupiter and Saturn flooded with new ice.

Note: Radiometric dating calculates an age for geologic materials by measuring the presence of a short-life radioactive element, e.g., carbon-14, or a long-life radioactive element plus its decay product, e.g., potassium-14/argon-40. The technique was first published in 1907 by Bertram Boltwood and is now the principal source of information about the absolute age of rocks and other geological features.

Q43. Out of the two sides of the moon (near side and far side), which side has more maria basins and why?

A43. The two hemispheres of the moon (near side and far side) have distinctly different appearances. The near side (the site visible from earth) is covered in multiple large maria while the far side (the site which is not visible from earth) has a densely cratered appearance with few maria. Only 1 per cent of the surface of the far side is covered by maria compared to 31.2 per cent on the near side. One commonly accepted explanation for this difference is related to a higher concentration of heat-producing elements on the near-side hemisphere, therefore having more volcanic eruptions. While the other factor is crustal thickness. The relatively low-density crust is thinner on the near side and has thus been ruptured by hotter material from the moon's interior at certain stages of the moon's history. The thicker crust on the far side has prevented this happening. Also, the thicker crust is less likely to have been completely ruptured by large impacts. The uneven crust and distribution of heat producing elements can be explained by the tidal lock of the moon: One side of the moon always feels a larger pull from earth's gravity than the other.

Q44. How have marias been named?

A44. The modern system of lunar nomenclature was introduced in 1651 by Giovanni Battista Riccioli. Riccioli's map of the moon was drawn by Francesco Maria Grimaldi. The International Astronomical Union (IAU) is the body responsible for assigning official names and designations to celestial bodies. Besides naming as Mare, the traditional nomenclature also includes Oceanus (ocean), lacus (lake), palus (marsh), and sinus (bay). The last three are smaller than maria but have the same nature and characteristics. As illustration, some of the names are:

Mare Tranquillitatis (Sea of Tranquillity), Lacus Veris (Lake of Spring), Sinus Amoris (Bay of Love), Palus Epidemiarum (Marsh of Epidemics).

Q45. What is the largest maria basin and how big is it?

A45. The largest maria basin on the moon is Oceanus Procellarum (Ocean of Storms). This is a massive region towards the western edge of the near side of the moon measuring approximately 2500 kilometres (1550 miles) by 770 kilometres (480 miles). Procellarum may be as young as 2 billion years. It can be seen from earth with unaided eyes if the phase of the moon is close to full.

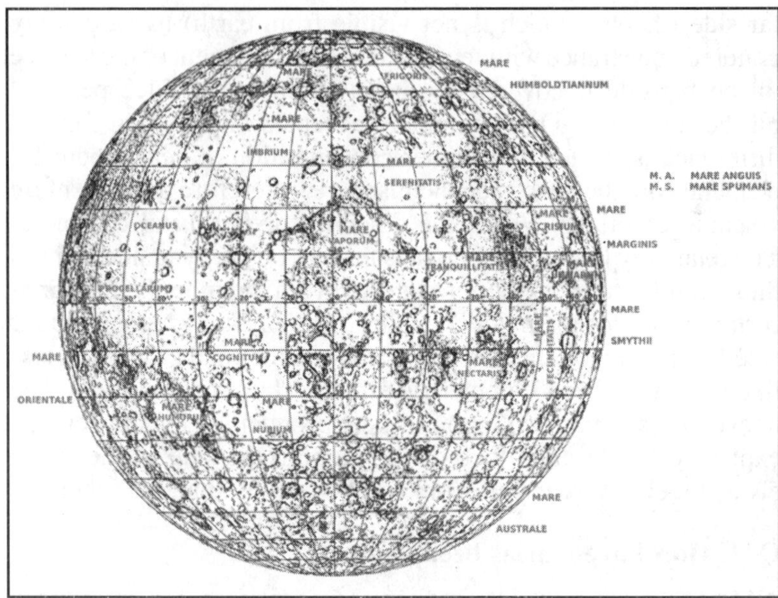

Moon-mare-near side.
Credits: Wikimedia Commons
Author: H. Seldon. October 2, 2009
Figure in public domain
https://commons.wikimedia.org/wiki/File:moon-mare-near_side.svg

Lighter Areas on the Moon's Surface (Highlands/Mountains)

Q46. What are lighter areas on the moon's surface?

A46. The lighter areas on the moon's surface are the lunar highlands. 83 per cent of the moon's crust consists of the lunar highlands. They are made of relatively low-density rock that solidified on the cooling moon like slag floating on the top of a smelter. The newborn moon was covered with an ocean of molten lava hundreds of kilometres deep. Gradually, the lava ocean cooled and solidified, forming the light-coloured crust. The highlands are composed predominantly of calcium-rich feldspar called anorthosite whereas the maria are basaltic, low in silica content, dark in colour, and comparatively rich in iron and magnesium. The highlands are older than the maria, and hence are more heavily cratered.

Q47. How do maria differ from highlands?

A47. Maria are darker, less cratered, smoother, younger, higher in iron, and generally lower in elevation than the highlands.

Maria are mostly composed of dark basalts, which form from rapid cooling of molten rock from massive lava flows. The Highlands rocks are largely Anorthosite, which is a kind of igneous rock that forms when lava cools more slowly than in the case of basalts.

Maria are smooth because when the lava flowed out it covered the rough surface and hardened smoothly. Since they formed, the maria have not been impacted as much as the older highlands.

Q48. Are highlands older than maria?

A48. More craters in the lunar highlands indicate the highlands are older than the mare areas. Different parts of the surface of the moon exhibit different amounts of cratering, an indicator of different ages. The oldest surfaces in the Solar System are characterized by maximal cratering density. Maria are younger and have not been hit as often. The lunar maria may have an average age of about 3.8 billion years as indicated by crater densities. Most highland samples have ages in the range 3.8–4.3 billion years.

Q49. Does the moon have mountains?

A49. Yes, the moon is home to some massive mountains. The mountains are referred to as massifs and known as terrae (singular terra, from the Latin for earth). Unlike the mountains on earth, the moon's mountains do not have any sharp folds in their ranges. The mountains have low, rounded profiles that resemble the oldest, most eroded mountains on earth. This is because there is no wind, water, or ice on the moon to carve them into cliffs and sharp peaks, the way we have seen them shaped on earth. Their smooth features are attributed to gradual erosion, mostly due to impact cratering from meteorites.

Q50. How were mountains formed on the moon?

A50. On earth, mountains are formed primarily by two different processes: tectonic action and volcanic action. Most are formed by tectonic action. Our earth has tectonic plates (also called lithospheric plates) which are a massive, irregularly shaped slab of solid rock. When these plates collide with each other, volumes of rock lift, slowly creating an elevated landform. Over millions of years, wind, water, and gravity erode these uplifted sections, wearing their surface to make the mountains we are familiar with today. Other mountains on earth are caused by volcanic action. Hot magma welling up from the depths of the earth to deposit itself on the surface as volcanoes.

The moon does not have tectonic plates. Earlier it was thought that virtually all its mountains are the result of impacts by asteroids in the distant past. Early in the moon's history, many gigantic asteroids in the solar system impacted the moon and formed craters and leaving their rims to form lunar mountain ranges. During impact events, mountains can appear not only on the border of the resulting crater, but also in its centre.

However, India's Chandrayaan-1 has revealed direct evidence on the formation of highlands on the moon by "lunar magma ocean hypothesis". The Indian mission through the moon Mineralogy Mapper instrument has offered proof that the moon's highlands were formed by the eruption of a hot liquid inside the moon's surface popularly known as magma. While it has been held that only the lowlands were formed by magma, this new evidence suggests even the highlands were formed by a similar process. The hot liquid, magma, seems to have flowed on to the surface and taken the form of lava. The rocky remains that floated to the top appear to have transformed into the moon's highlands or mountains. Investigators report that the entire Inner Rook Mountain range (a ring-shaped mountain range that lies along the western limb of the moon, crossing over to the far side) is of this character. Japan Aerospace Exploration Agency's imager on Japan's Kaguya spacecraft is reporting similar findings in 70 impact craters around the moon. Corroboration from other space agencies indicates similar results that enhance science data from Chandrayaan-1.

Crater Schrödinger on the moon, in southern part of the far side (The 312 kilometres wide Schrödinger crater with its beautiful mountain ring)
Credits: Source JMARS
Author: NASA (image by Lunar Reconnaissance Orbiter)
Date: composed from images, taken on different dates
Wikimedia Commons
https://commons.wikimedia.org/wiki/File:Schr%C3%B6dinger_(LRO)_500_km.png
Figure in public domain

Q51. How are the moon's highlands named?

A51. There are two types of lighter-coloured highlands on the moon: mons (Latin word for "mountain"), which are isolated single mountains and montes (plural of mons), which are mountain ranges. In planetary nomenclature, these terms are capitalized and stands before the proper given names for lunar mountains, example Montes Taurus, Mons Hadley etc. The term tholus ("dome") is used for names of smaller uplands, and the term colles ("hills") in names of groups of still smaller.

The International Astronomical Union (IAU) approved the first names of extraterrestrial mountains in 1935. As of May 2015, 262 extraterrestrial mountains and mountain systems have official names, out of which 48 are on the moon.

Q52. What is the highest mountain on the moon and how does it compare with earth's highest mountain?

A52. Let us first consider the earth. Measurement of mountains height on earth is looked at with a reference frame. There are three reference frames:

1. Height above sea level
2. Base to peak height
3. Height above centre of the earth

Mount Everest (Nepal and Tibet) is the tallest mountain from sea level, Mauna Kea (Hawaii, USA) is the tallest from base to peak, and Chimborazo (Ecuador, South America) is the farthest point from the centre of the earth.

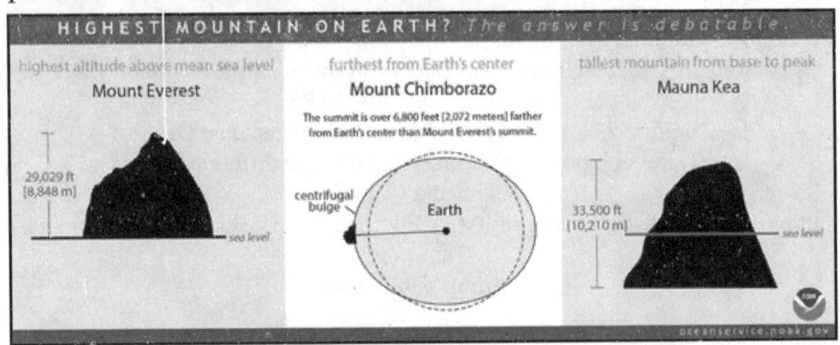

Highest mountain on earth
Credits: National Oceanic and Atmospheric Administration (NOAA). Taken from website on September 7, 2020
Figure in public domain
https://oceanservice.noaa.gov/facts/highestpoint.html#transcript

Now coming to the moon. We consider 'base to peak height' as a reference frame. The lunar Mon Huygens located in Montes Ampere (near South Pole Aitken basin) is the highest mountain at about 5500 metres (18045 feet), more than half the height of mount Everest 8850 metres (29000 ft). The mountain was named after the Dutch astronomer, mathematician, and physician Christiaan Huygens.

There are several major differences between Mt. Everest and Mon Huygens. Mt Everest is a relatively new feature on the earth. It was formed as tectonic plates collided and pushed up to astonishing heights what was once seafloor, over the course of about 60 million years. The lunar high point is very ancient and is formed by the impact that created Mare Imbrium in a matter of minutes, more than 4 billion years ago. Another key difference between the two highest points is slope. The slopes of Mt Everest are very steep, while on the moon it is not.

But even though it is the moon's highest mountain, Mons Huygens is not the moon's highest point. That would be the Selenean summit, which is 10786 metres (35387 ft) above the lunar mean (moon's centre). It is nearly twenty per cent taller than earth's relative highest point Everest. Mons Huygens is still the highest mountain though, because the Selenean summit's slopes only average 3 degrees, which isn't steep enough for it to be considered a mountain.

Q53. Which regions of the moon have the highest and the lowest elevations?

A53. Highest Point on the Moon

Selenean summit refers to the "highest" point on the moon, notionally like Mount Everest on the earth. At some 10,786 metres (35,387 feet) above the lunar mean, it is nearly twenty per cent 'taller' than earth's relative highest point, Everest. The summit is located along the north-eastern rim of Engel'gardt crater. The summit is located on the far side of the moon relative to earth.

Lowest Point on the Moon

The bottom of a small unnamed crater within Antoniadi crater (on the floor of the South Pole - Atkin) was measured by the Japanese SELENE satellite to be the lowest point on the moon. The point is 9178 metres (30112 feet) below the average lunar mean (geoid). The measurement was confirmed by the Chinese Chang'e 1 orbiter. The lowest point has been unofficially named "Point Trieste".

Craters

Q54. Why does the moon have craters?

A54. Craters (from the Greek word for vessel) are the most dominant lunar surface features. They record the history of the Solar System. The main reason the moon has craters is because it gets hit by objects, small and big pieces of rocks that come from outer space. These are pieces of asteroids, comets, meteorites that are flying around in the solar system. When they hit the surface, there is an impact and craters are generated. It is the speed and angle of approach of an impactor which determines how big a crater it will leave behind. Some have suggested that the moon also has craters because in its past it had volcanoes which can create craters. However, evidence suggest that most of the craters are by impacts.

Q55. How are craters made on the moon?

A55. When an impactor strikes the solid surface of the moon (or a planet), a shock wave is created which spreads out from the site of the impact. The shock wave fractures the rock and excavates a large cavity, much larger than the impactor itself. Thus, a crater gets formed. The impact sprays ejecta material out in all directions. The impactor is shattered into small pieces and may melt or vaporize. Sometimes the force of the impact is great enough to melt some of the local rock. If an impactor is large enough, some of the material pushed toward the edges of the crater will slump back towards the centre and the rock beneath the crater will rebound, or push back up, creating a central peak in the crater. The edges of these larger craters also may slump, creating terraces that step down into the crater.

In recent times, one crater formation was visible to the naked eye. On March 17, 2013, an object the size of a small boulder (40 kg) hit the surface in Mare Imbrium at about 40000 miles per hour. ASA's Lunar Reconnaissance Orbiter (LRO) acquired images of the lunar surface before and after the recorded explosion occurred on the surface. The crater itself is small, measuring 18.8 metres (61.7 feet) in diameter, but its influence is large. Debris excavated by the sudden release of energy flew for hundreds of metres. More than 200 related surficial changes up to 30 kilometres (19 miles) away were noted.

Q56. What are some of the important features of the moon craters?

A56. Craters typically have the following features:

1. A surrounding area with materials splashed out of the ground when the crater was formed.
2. Raised rim, consisting of materials ejected but landing very close by.
3. Crater wall, the downward-sloping portion of the crater.
4. Crater floor, a more or less smooth, flat area, which as it ages accumulates small craters of its own.
5. Central peak (found only in some craters with a diameter exceeding 26 kilometres (16 mi)): this is generally a splash effect caused by the heat energy created by impacting object and melting some lunar material.

Tycho Central Peak (sunrise view of Tycho crater captured by LRO spacecraft)
Tycho crater is ~82 km (51 miles) in diameter. The summit of the central peak is 2 km (6562 ft) above the crater floor, and the crater floor is about 4700 m (15,420 ft) below the rim.
Credits: Wikimedia Commons, the free media repository
Author: NASA / GSFC / Arizona State University. June 10, 2011
https://commons.wikimedia.org/wiki/File:LRO_Tycho_Central_Peak_0.50.jpg
Figure in Public Domain

Q57. How are the moon's craters named?

A57. Since 1919, assignment of these names is regulated by the International Astronomical Union (IAU). Some of the methodologies used for naming craters are given below:

Small craters of special interest, for example, visited by lunar missions, receive human first names like Robert José Louise, etc.

One of the biggest lunar craters, Apollo, is named after Apollo missions. Many smaller craters inside and near it bear the names of deceased American astronauts, and many craters inside and near Mare Moscoviense bear the names of deceased Soviet cosmonauts.

In 1970 twelve craters were named after twelve living astronauts (6 Soviet and 6 American).

Satellite craters' names consist of the name of a nearby named crater and a capital letter, for example, Copernicus A, Copernicus B, Copernicus C and so on.

Lunar crater chains are usually named after a nearby crater. Their Latin names contain the word Catena ("chain"). For example, Catena Davy is situated near the crater Davy.

Q58. How many craters are on the moon?

A58. Craters occupy most of the moon's surface. The number seems countless. Scientists estimate that there are around 1 million craters larger than half a mile across and over half a billion that are larger than 10 metres wide. But these are estimates. Nobody knows the actual count. Now only a few of them have been named and dated. The International Astronomical Union (IAU) currently recognizes 9,137 craters, of which 1,675 have been dated.

To effectively identify craters and estimate their age, a team of researchers from Jilin University applied a transfer learning method. By combining the data collected by China's Chang'e 1 and Chang'e 2 lunar probes, researchers identified 109,956 new impact craters. They also estimated the ages of 18,996 newly detected craters larger than 8 kilometres in diameter. According to these researchers, the lunar crater database is of great value to scientific research on the moon.

Identifying smaller craters has been a problem. Based on high resolution images of the moon's surface, researchers from the Curtin University in Australia have developed a new algorithm to reliably identify craters on the moon that are less than a kilometre wide. The researchers indicate that with the growing availability of high-resolution images, automated crater counting will be crucial to determining ages of smaller features on the moon's surface and other planetary bodies.

Reference

Yang, C., Zhao, H., Bruzzone, L. et al. Lunar impact crater identification and age estimation with Chang'E data by deep and transfer learning. Nat Commun 11, 6358 (2020). https://doi.org/10.1038/s41467-020-20215-y

Q59. Why does the moon's far side have more craters than its near side?

A59. When the Soviet probe Luna 3 sent back the first shots of the far side of the moon (year 1959), they showed that the far side has more craters than the near side.

The reason for more impact craters on the far side of the moon is that the near side has a much thinner crust which has allowed volcanoes to erupt and fill in ancient large impact craters. The reason for the different crustal thickness of both the sides is proposed by a new hypothesis. After the impact that created the moon, it got tidally locked to the earth in only a few days. Because the earth was still hot (2,500°C) from the residual heat of impact, it cooked the near side but let the far side cool more rapidly, thus forming a thicker crust than the near side. The large lava flows in the near side have covered craters that were formed early in the moon's history. It is likely that each side of the moon has received equal numbers of impacts, but the resurfacing by lava results in fewer craters visible on the near side than the far side, even though both sides have received the same number of impacts.

Some people argue that the reason the far side of the moon has more impact craters than the near side is because the earth shields the near side from asteroids impacts. But that is not the real reason. The earth partially shields the near side (only about 0.01percent) of the moon from incoming asteroids, that is not a large enough effect to influence crater densities.

Q60. Why are the highlands more cratered than the maria?

A60. The highlands are heavily cratered because they were formed early in lunar history (between 4.1 and 4.4 billion years ago) bearing the scars of all those billions of years of impacts by interplanetary debris. On the contrary, the maria were formed after large impacts from meteors carved out basins in the lunar crust. When the moon was volcanic, magma seeped to the surface, filled the basins, and eventually hardened, resulting in the relatively smooth flat areas seen today. Scientists can tell this happened recently, in geological terms.

Q61. Which is the largest crater in the moon?

A61. The South Pole-Aitken basin is the oldest and deepest crater on the moon, and one of the largest craters in the entire solar system. It is 2,500 kilometres (1550 miles) wide and 13 km (8 miles) deep. It is in the far side of the moon.

Q62. Why does the moon have so many craters while the earth has so few?

A62. Both the earth and the moon have been hit several times throughout their long 4.5-billion-year history. As a matter of fact, an asteroid or meteor is more likely to fall towards the earth than the moon because earth's stronger gravity attracts more space debris. But we can see many thousands of craters on the moon, and we only know of about 190 (Wikipedia) on earth.

The reason is that the earth has processes that can erase almost all evidence of past impacts. The moon does not. Pretty much any tiny dent made on the moon's surface is going to stay there.

The following three processes keep the earth's surface crater free.

1. Erosion on earth due to weather, water, and plants. The moon has almost no erosion because it has no atmosphere, no wind, no weather, and no plants. Almost nothing can remove marks on its surface once they are made except another impact.
2. Tectonics that cause the earth's surface to form new rocks, get rid of old rocks, and shift around over millions of years. Because of tectonics, the surface of earth is recycled many times throughout its long history. As a result, very few rocks on earth are as old as the rocks on the moon. Most of the earth is covered by oceans, and because of plate tectonics, the ocean floor is constantly being reworked. The moon does not have tectonics. Approximately 80 per

cent of earth's surface is less than 200 million years old, while over 99 per cent of the moon's surface is more than 3 billion years old. Essentially, the moon's surface has not been modified since early in its history, so most of its craters are still visible.

3. Volcanic flows can cover up impact craters. This is a major way impact craters get covered up elsewhere in our solar system. The moon once had large volcanic flows way in the past that did cover up many of the bigger earlier impacts, but it has been without volcanism for around three billion years.

Q63. Why are the moon craters larger in the "near side" of the moon than in the "far side"?

A63. Planetary scientists have shown that the moon's near side have significantly larger craters than the far side. Volcanic activity, that occurred about 4-billion years ago, known as the Late Heavy Bombardment (also Lunar Cataclysm), resulted in the crust and upper mantle being hotter on the moon's near side than on its farside. This is because after the impact that created the moon, it got tidally locked to the earth in only a few days. Because the earth was still hot (2,500°C) from the residual heat of impact, it cooked the near side more and for a longer time than the far side which was not facing the earth. Variations in temperature within the moon's crust have a large effect on the final size of craters. This heat made the moon's nearside geology more susceptible to expanding readily outward following an asteroid impact. On the moon's cooler farside, crust beneath the crater rim was prone to collapsing inward resulting in a diameter that is smaller than the near side. Simulations has confirmed this.

Reference

Gonzalez, Robbie. What created such enormous craters on the moon's nearside? GIZMODO. November 7, 2013 https://io9.gizmodo.com/why-are-craters-bigger-on-the-moons-nearside-1460257773

Rocks on the Moon

Q64. How were rocks formed on the moon?

A64. Almost all the rocks at the lunar surface are formed when magma cools and hardens, known as igneous rocks. By contrast, the most prevalent rocks exposed on the earth's surface are sedimentary which formed by the collection, compaction, and cementation of mineral grains, rock fragments, and sand that are moved by wind, water, or ice to the site of deposition. The two most common kinds of rocks on the moon are basalts (rich in magnesium and iron) and anorthosites (mostly plagioclase feldspar)

Astronaut Harrison "Jack" Schmitt at Tracy Rock on the lunar surface photographed during the Apollo 17 mission
Credits: Wikimedia Commons, photographed by astronaut Eugene A. Cernan
https://commons.wikimedia.org/wiki/File:Apollo_17_AS17-140-21497.jpg
Author: NASA
DATE: December 13, 1972
Figure in Public Domain

Q65. Why does the moon have big rocks?

A65. The moon is made up of various kinds of rocks, fairly like the rocks on earth. But on earth, we have wind and rain that wear the rocks down into sand and dirt. There is no air or wind on the moon, so the rocks do not get worn down as they do on the earth.

Tracy's Rock (known as Split Rock or the Station 6 Boulder in the scientific literature) is a large boulder on the moon which was visited by the Apollo 17 crew on December 13, 1972.

Q66. What kinds of rocks are there on the moon?

A66. Moon rocks fall into two main categories: basalt - those found in the maria, and anthrosite—those in the lunar highlands (terrae).

Basalt: The Mare Rock

The lunar maria are composed of rock basalt which is like the kind of volcanic rock found on earth. The lunar basalts are rich in iron and magnesium. Maria rocks contain glassy structures indicating rapid cooling. However, unlike earth basalts, the lunar samples does not contain water and a lower percentage of volatile. Basalt is not distributed uniformly over the moon. Nearly 26 per cent of the near side of the moon is basalt and only 2 per cent of the far side is basalt.

Anorthosite: Highland Rock

The highlands, the large portions of the lunar crust, consist of anorthosite (a category of Feldspar), a calcium-rich white rock and materials thrown out during the creation of the impact basins. Over 80 per cent of the moon seems to be light coloured feldspar, which is also the most common rock on earth. This supports a theory that at one time the entire surface of the moon was molten, just one huge magma ocean 480-650 kilometres (300-400 miles) deep. As it cooled, the lighter minerals, like feldspar, floated to the top and formed anorthosite. Heavier minerals sank and produced the denser interior of the moon.

There is one mystery though. The age of the highlands doesn't match how long it would have taken for the moon's magma ocean to cool. A new study suggests that the interior of the moon was more of a frozen slushy magma than a viscous fluid, explaining the lunar highlands.

Q67. What are breccias and how did they form on the moon?

A67. The word "breccia" has an Italian origin which means "stone made of cemented gravel." Most of the rocks on the moon are breccias, which are hard, solid rocks that have broken up and formed back together in different combinations. Lunar breccias are rocks produced by the smashing, melting, and mixing of the lunar surface materials by large and small meteoric impacts. Evidence of this process can be seen in the countless craters of various sizes which cover the moon. Some rock fragments found in breccias are pieces of more ancient breccias. Repeated impacts have smashed the older rock and re-fused it with more recently formed breccia. As many as four generations of breccia have been found in a single lunar rock. These are known as “breccia in breccia”.

A sample of Apollo 11 breccia is a glassy material produced by the shock of a meteoric impact. Since the glass is not uniform in composition, it strongly indicates that the glass was formed by shock melting. This ellipsoidal lunar glass particle contains numerous tiny spheres of nickel-iron.

A sample from Apollo 17 Breccia is lithified mature soil (Lithification refers to complex physical, chemical, or biological processes whereby unconsolidated material (e.g., sand, silt, and mud) becomes converted to solid rock). The sample consists of fragments of glass, minerals, and rock cemented together in a glassy matrix. The materials which compose this sample have been determined to be 4.53 billion years old

Breccias are not unique to the moon. Volcanic areas on earth have many breccias, as well.

Q68. How old are the moon rocks?

A68. The moon rocks show that the lunar surface is very ancient. Its rocks are older than most of the rocks now found on earth.

This way the moon has preserved a record of the earliest history of the formation and development of the planets. They range in age from about 3.16 billion years old for the basaltic samples derived from the lunar maria, up to about 4.5 billion years old for rocks derived from the highlands. The Apollo missions returned samples of ancient lunar rocks which are about 4.5 billion years old, indicating that parts of the moon's crust solidified soon after the moon formed.

Q69. Why are the rocks on the moon older than those on earth?

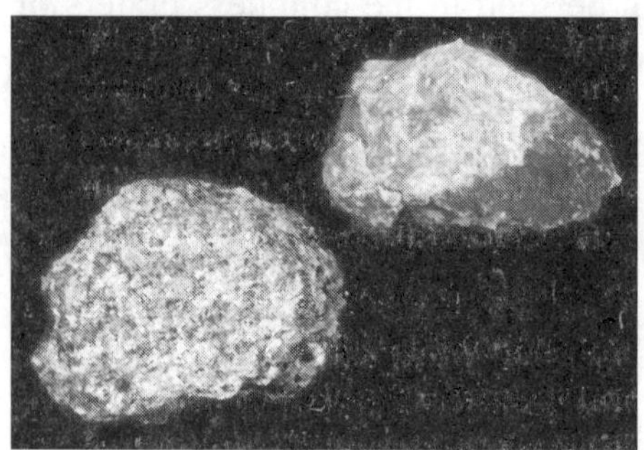

Oldest Moon Rocks
These rocks are about 4.5 billion years old, indicating that parts of the moon's crust solidified soon after the moon formed.
Credits: courtesy of NASA Johnson Space Centre.
https://www.lpi.usra.edu/education/timeline/gallery/slide_15.html

A69. The earth has tectonic plate movements, volcanic activity, and weathering from air and water (erosion). The moon does not.

The movement of tectonic plates in the earth's lithosphere (the rigid outer part of the earth, consisting of the crust and upper mantle) and the eruption of volcanoes effectively "recycle" the rock on the surface, consuming or burying the older rocks while creating new ones. Over time, some of earth's oldest rocks have been eroded by the action of air and water.

In contrast the moon has remained essentially static for billions of years. So, we see old surfaces on the moon, heavily populated by craters formed from massive meteoric impacts early in the history of our Solar System. These bodies are valuable as an enduring record of that history.

Moon Dust (Regolith)

Q70. What is lunar dust (regolith) made of?

A70. The bulk of the lunar dust (known as regolith) is a fine grey soil covering almost the entire lunar surface with a density of about 1.5 g/cm3. It also includes breccia and rock fragments from the local bedrock. This regolith has formed over the last 4.6 billion years from the impact of large and small meteoroids, from the steady bombardment of micrometeoroids and from solar and galactic charged particles breaking down surface rocks.

The regolith is generally from 4 to 5 metres thick in mare areas and from 10 to 15 metres in the older highland regions.

When the plans were put together for the Apollo missions, some scientists were concerned that the lunar regolith would be too light and powdery to support the weight of the lunar lander. Instead of landing on the surface, they were worried that the lander would just sink down into it like a snowbank. However, landings performed by robotic Surveyor spacecraft showed that the lunar soil was firm enough to support a spacecraft, and astronauts later explained that the surface of the moon felt very firm beneath their feet.

China's Chang'e-4 rover has been crawling around on the moon's far side, near the south pole, and has made a surprising discovery. The regolith on the far side is stickier than the near side. Engineers have noticed more dirt clumps attaching to the rover's wheels than seen on previous nearside rovers.

Note: The term regolith is a combination of two Greek words: rhegos 'blanket', and lithos 'rock'. Regolith is a blanket of unconsolidated, loose, heterogeneous superficial deposits covering solid rock. It includes dust, broken rocks, and other related materials and is present on earth, the moon, Mars, some asteroids, and other terrestrial planets and moons.

Q71. What other terms are being used with reference to lunar regolith?

A71. Astronomers/Scientists use the following terms:

Megaregolith: Below the regolith is a region of blocky and fractured bedrock created by larger impacts, which is often referred to as the megaregolith.

Lunar Soil: The term lunar soil is typically referring to the finer fraction of regolith which is composed of grains one centimeter in diameter or less. About half the weight of a lunar soil is less than 60 to 80 microns in size.

Lunar dust: This is a finer material than lunar soil, the fraction which is less than 30 micrometers in diameter. The depth of the overlying dust is only a couple of centimeters thick and quite firm, so there is no fear of sinking.

Lunar dust (Regolith) is powdery, and sticky as can be seen in this photograph. Apollo 11 Lunar Module Pilot Buzz Aldrin's boot print. Aldrin photographed this boot print about an hour into their lunar extra-vehicular activity on July 20, 1969, as part of investigations into the soil mechanics of the lunar surface.
Credits: Source NASA, Author NASA/Buzz Aldrin
https://en.wikipedia.org/wiki/File:Apollo_11_bootprint.jpg
Figure in public domain

Agglutinates: Since the moon lacks an atmosphere, the upper few millimeters of the regolith is exposed to the bombardment of micrometeorites and to solar wind radiation. The extensive bombardment by micrometeorites, which continues even today, breaks up soil particles and melts portions of the soil. The melt, mixed with sand sized rock pieces, forms irregular clusters called agglutinates. At the same time, the solar wind implants large quantities of Hydrogen and Helium and trace amounts of other elements.

Q72. Is the moon dust dangerous?

A72. The harmful properties of lunar dust are not yet well understood. It is expected that exposure to lunar dust will result in greater risks to health both from acute and chronic exposure. This is because lunar dust is more chemically reactive and has larger surface areas composed of sharper uneven edges.

Once astronauts reached the surface, they reported that the fine moon dust stuck to their spacesuits. The astronauts also claimed that it got into their eyes, making them red; and worse, even got into their lungs, giving them coughs. Lunar dust is very abrasive and has been noted for its ability to wear down spacesuits and electronics.

A 2005 NASA study ranked "moon dust" as the number 1 challenge. They listed 20 risks that required further study in the areas of its mechanical properties, corrosiveness, grittiness, and effect on electrical systems. The dust found on the lunar surface could cause harmful effects on any human outpost technology and crew members. Some of the harmful effects could be:

1. Darkening of surfaces, leading to a considerable increase in radiative heat transfer.
2. Abrasive nature of the dust particles may rub and abrade surfaces through friction.
3. Negative effect on coatings used on gaskets to seal equipment from space, optical lenses, solar panels, and windows as well as wiring.
4. Possible damage to an astronaut's lungs, nervous, and cardiovascular systems.

Q73. How lunar regolith can be used for human purposes?

A73. Scientists are planning long-duration missions to the moon. There is no way lunar explorers will be able to bring everything they need to survive on the moon. It may be possible to supplement some of the materials needed from resources on the moon itself. Regolith is one of such resources. The following are some of the areas in which experiments are in progress:

1. **Regolith as a building material:** Scientists have been thinking of ways that we could potentially use the regolith as a kind of building material. A research team from Russia's Skoltech Center for Design, Manufacturing and Materials (CDMM) has published a review paper which summarizes the worldwide progress in 3D printing techniques using lunar regolith and categorizes them based on different applications in building future moon habitats. The team's own focus is on manufacturing high-strength ceramics, which they started building using lunar soil simulant from NASA.
2. **Source of Oxygen:** Approximately 45 per cent of the weight of regolith is oxygen, but it is bonded to materials such as iron and titanium. To utilize both the oxygen and the materials it's bonded to, they must be separated. A British company Metalysis, with support from the European Space Agency (ESA), has begun testing a technique to judge its potential effectiveness on the moon.
3. **Germination and plant growth in lunar regolith:** Scientists at Wageningen University and Research, Netherlands have reported on the first large-scale controlled experiment to investigate the possibility of growing plants in Mars and moon soil simulants. Their results show that plants can germinate and grow on both Martian and moon soil simulant for a period of 50 days without any addition of nutrients. However, many questions remain about the simulants' water carrying capacity and other physical characteristics and whether the simulants are representative of the real soils.
4. **Helium-3 mining on the lunar surface:** The moon has been bombarded with large quantities of Helium-3 by the solar wind. In 1986, scientists at the Institute of Fusion Technology at the University of Wisconsin estimated that the lunar regolith, contains one million tons of helium-3 (He3),

a material that could be used as fuel to produce energy by nuclear fusion since it is not radioactive and would not produce dangerous waste products. Many nations including India, China, Russia, and the USA consider lunar helium-3 a potential economic resource to be mined.

References

1. *Tomaswick, Andy. 3-D Printing on the Moon From Regolith to Paste to Useful Objects and Structures. Universe Today. January 24, 2021.* https://www.universetoday.com/149771/3-d-printing-on-the-moon-from-regolith-to-paste-to-useful-objects-and-structures/

2. *Tomaswick, Andy. Figuring Out How to Breathe the Moon's Regolith. Universe Today. November 30, 2020.* https://www.universetoday.com/148990/figuring-out-how-to-breathe-the-moons-regolith/

3. *Wamelink, G.W.W. Frissel, J.Y. Krijnen, W.H.J. Verwoert, M.R. Goedhart, P.W. Can Plants Grow on Mars and the Moon: A Growth Experiment on Mars and Moon Soil Simulants. Wageningen University & Research, Netherlands. August 27, 2014* https://doi.org/10.1371/journal.pone.0103138

Q74. What are moon dust fountains?

A74. There is some evidence that the moon has a layer of moving dust particles constantly flying up and falling back to the moon's surface, giving rise to a dust atmosphere, known as moon fountain. The term "moon fountain" (like drinking water fountain) has been used to describe this effect. This is caused by electrostatic levitation. Moon's air-free and barren surface is constantly being bombarded by charged particles from the sun. Some loose grains of dust, acquiring similar charge to the nearby surface, levitates up against gravity up to several feet. In the early 1960s before Apollo 11, several early Surveyor spacecraft that soft-landed

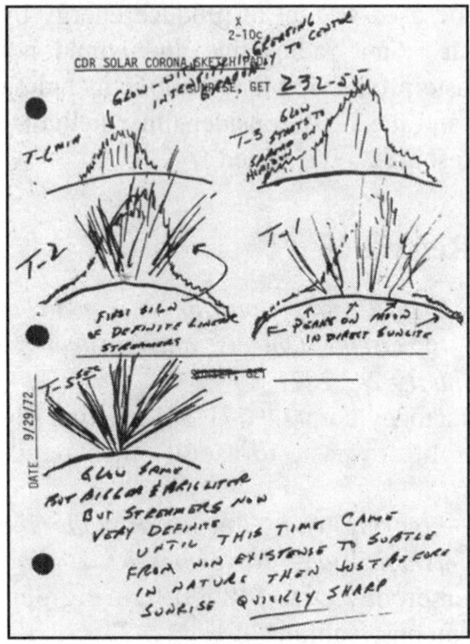

Lunar "twilight rays" sketched by Apollo 17 astronauts
Credits: Wikimedia Commons
Author: Apollo 17 astronauts
Date: September 29, 1972
This figure is in public domain
https://commons.wikimedia.org/wiki/File:Apollo_17_twilight_ray_sketch.jpg

on the moon returned photographs showing an unmistakable twilight glow low over the lunar horizon persisting after the sun had set. Moreover, the distant horizon between land and sky did not look razor-sharp, as would have been expected in a vacuum where there was no atmospheric haze. Apollo 17 astronauts orbiting the moon in 1972 repeatedly saw and sketched what they variously called "bands," "streamers" or "twilight rays" for about 10 seconds before lunar sunrise or lunar sunset. Such rays were also reported by astronauts aboard Apollo 8, 10, and 15.

Electrostatic levitation, though offers a plausible explanation, is not strong enough to create and sustain the mysterious, glowing clouds. Their origin remains something of mystery.

Reference

Moon Fountains. Science Mission Directorate (nasa.gov). March 30, 2005
https://science.nasa.gov/science-news/science-at-nasa/2005/30mar_moonfountains

Other Features on the Moon Surface

Lunar Rays

Q75. What are the lunar rays?

A75. Lunar rays, also known as lunar crater rays, are bright streaks of pulverized rock radiating outward from many impact craters. They look somewhat like many thin spokes coming from the hub of a wheel. These rays are among the most prominent lunar features. When the sun is high in the lunar sky the bright lunar rays blaze into view. The rays may extend to a distance of a few crater radii, but rays from some of the craters like the Copernicus and Tycho craters, can be seen halfway around the moon. Rays appear to be distributed randomly across the lunar disc although those on the dark maria tend to show up most dramatically because of the contrast effect.

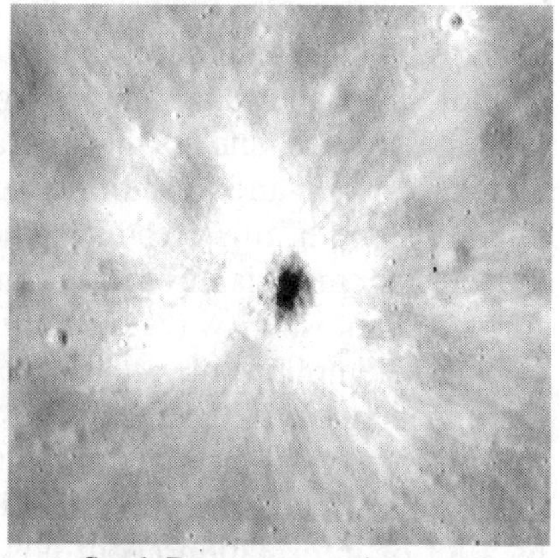

South Ray crater on the moon
AS16-P-4618 ASU (Baby Ray crater is in upper right)
Credits: Wikimedia Commons
Author: James Stuby based on NASA Apollo 16 Panoramic camera image
Immediate source: Arizona State University, Apollo Browse Gallery AS16-P-4618
Date: 2015, Original 1972
https://commons.wikimedia.org/wiki/File:South_Ray_crater_AS16-P-4618_ASU.jpg
Figure in Public Domain

Ray systems have been identified on the moon, earth (Kamil Crater), Mercury, and some moons of the outer planets.

The picture below is of South Ray crater (approximately 700 m in diameter and approximately 120 m deep) showing a bright system of rays of ejecta. The original picture is an Apollo 16 Panoramic camera image cropped by James Stuby of Arizona State University.

Q76. How did lunar rays form?

A76. The physical nature of lunar rays has historically been a subject of speculation. American geologist Eugene Shoemaker suggested during the 1960s that the rays were the result of fragmented ejecta material during meteoric and asteroidal impacts, which then settle onto the surface. According to one hypothesis, Sabuwala, and others have suggested that crater rays are formed due to the interaction of a shockwave that is generated when an impactor strikes the surface. If the surface is initially smooth, this creates a smooth shockwave that disperses the ejected material evenly — no rays. But with an uneven surface, this shockwave becomes asymmetrical, causing the ejecta to clump in some areas and form rays that spread out from the impact zone. Their results also suggest that crater rays have information about the sizes of the rocks that hit the moon to form the craters. If we can extract that information, then it would give us a lot of information about how our solar system formed and what it was like in its early years.

However, more observational data is still needed to test the theories of the origin and nature of these rays. The Association of Lunar and Planetary Observers (ALPO), along with several other organizations have embarked upon a study of these lunar rays.

Reference

Sabuwala, Tapan. Butcher, Christian. Gioia, Gustavo and Chakraborty, Pinaki. Ray Systems in Granular Cratering. PHYSICAL REVIEW LETTERS 120, 264501. June 27, 2018
https://journals.aps.org/prl/pdf/10.1103/PhysRevLett.120.264501

Lunar Valleys / Rilles

Q77. What are lunar valleys/rilles and how are they formed?

A77. Rille is typically used to describe any of the long, narrow depressions in the surface of the moon that resemble channels.

The Latin term is rima, plural rimae. They are sometimes labelled as valleys, cracks, or clefts. They are difficult to find with the naked eye. Typically, a rille can be up to several kilometres wide and hundreds of kilometres in length. Valleys are unusual because other features such as craters, or rays intersect them.

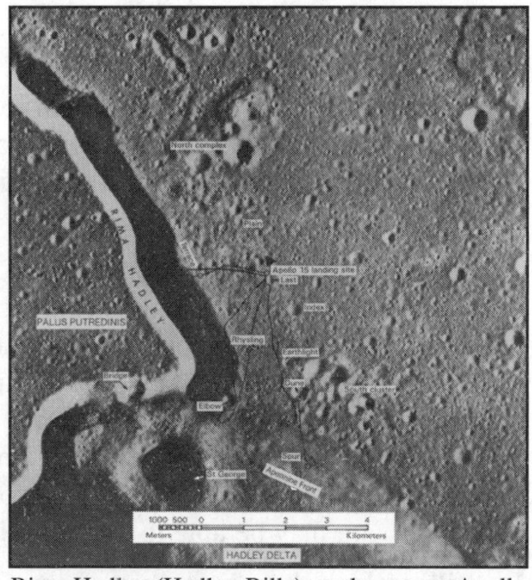

Rima Hadley (Hadley Rille) on the moon. Apollo 15 landed about 2 kilometres from the rim of Hadley Rille and explored it in detail.
Credit: NASA, Wikimedia commons
https://commons.wikimedia.org/wiki/File:Apollo_15_actual_traverses_and_sample_stations.jpg
https://www.hq.nasa.gov/alsj/a15/a15_lpi_trvrsmap.gif
Figure in public domain

Precise formation mechanisms of rilles have yet to be determined. They are thought to have been formed by ancient lava flows, collapsed lava tubes or geological faults. When lava flowed on the moon, the top and sides of these tubes solidified first while lava flowed underneath. When this lava moved away or solidified, the top or roof of the tube collapsed, leaving these rilles.

Q78. What are the different types of rilles?

A78. Three types of rille are found on the lunar surface:

Sinuous rilles resemble winding river valleys on earth. Rima Hadley, a sinuous rille, was visited by the Apollo 15 astronauts in 1971. It is interpreted as one of the most defined sinuous rilles on the lunar surface. It typically ranges in depth between 600 and

900 feet (180 and 270 m) but is approximately 1,200 feet (370 m) deep at the Apollo 15 landing site. The rille has a cumulative length of about 80 miles (130 km) and an average width of about 0.75 miles (1.21 km).

Arcuate rilles have a smooth curve and are found on the edges of the dark lunar maria. These are found all over the moon. Rima Sulpicius Gallus is a clear example in southwestern Mare Serenitatis.

Straight rilles are flat-floored and follow long, linear paths. These are believed to be like ditches/trenches (grabens), sections of the crust that have sunk between two parallel faults. These can be readily identified when they pass through craters or mountain ranges. The largest rille on the moon is a straight rille known as Rima Sirsalis. It extends about 400 km across the lunar near side in Oceanus Procellarum, the Ocean of Storms. On a good or clear night you may be able to find it with a telescope.

Rilles which show more than one structure are termed hybrid rilles.

Lunar lava tubes

Q79. What are the lunar lava tubes and how are they formed?

A79. Lunar lava tubes (also known as pyroducts) are lava tubes on the moon formed during the eruption of lava flows. Scientists believe that these are caves created by lava as it flows from a volcanic vent beneath a hardened surface. When the surface of a lava flow cools, it hardens, and the lava can channel beneath the surface in a tube-shaped passage. Once the flow of lava diminishes, the tube may drain, forming a hollow void. Lunar lava tubes are generally formed on sloped surfaces and are typically found along the boundaries between lunar mares and highland regions. The existence of a lava tube is sometimes revealed by the presence of a "skylight", a place in which the roof of the tube has collapsed, leaving a circular hole that can be observed by lunar orbiters.

The ISRO Chandrayaan-1 orbiter imaged a lunar rille formed by an ancient lunar lava flow with an uncollapsed segment indicating the likely presence of a lava tube near the lunar equator. The tunnel measures about 2 km (1.2 mi) in length and 360 m (1,180 ft) in width.

On earth, these lava tubes can be found in Hawaii, the Canary Islands, Australia, and Iceland, California, Washington-Oregon, the Deccan flats of southern India, in northeast Siberia, and elsewhere. They range in diameter from 33 to 98 feet while Lunar caves could reach 1 kilometre (3280 feet) in diameter and could be longer than 25 miles, enough space to host a human colony. Even the skylight openings can be as wide as a football field.

According to Riccardo Pozzobon, a geoscientist at the University of Padova, Italy, a lava tube on the moon could easily contain a small city within its walls. The sheer scale of these extraterrestrial lava tubes is likely a result of low lunar gravity, as well as differences in how volcanoes operated on the moon compared to earth.

Reference

Letzter, Rafi. Entire cities could fit inside the moon's monstrous lava tubes. Live Science. August 15, 2020 https://www.livescience.com/lava-tubes-mars-and-moon-habitable.html

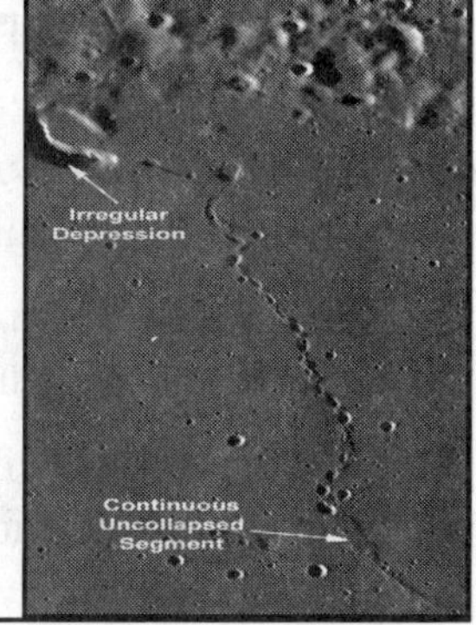

Sinuous chain of collapse pits transitioning into a continuous uncollapsed segment of a lunar lava tube. The chain is about 50 km long and is located at the transition from Mare Imbrium to Oceanus Procellarum, northwest from the crater Gruithuisen.
Credits: Wikimedia Commons, NASA/GSFC/ Arizona State University. Source: http://www.nasa.gov/mission_pages/LRO/multimedia/lroimages/lroc-20110217-chain.html February 27, 2014
Figure in Public Domain

Q80. How do lunar lava tubes differ from tubes found on earth?

A80. Some of the differences are:

(1) Formation of lava tubes on the moon are very ancient events on the order of 3.5-3.8 billion years ago. Surviving lava tubes on earth are all much younger than that, some only thousands of years old.

(2) Due to the lower gravity on the moon, lunar lava tubes are bigger than the earth's, in width, ceiling height, and total length.

(3) Compared to lava tubes on earth that get eroded by earthquakes, plate tectonics and other processes, lunar lava tubes have been largely untouched for 3 to 4 billion years. Lunar lava tubes have never been exposed to air or water (unless a comet happen to pierce the ceiling). The temperature will be colder than earth's.

Q81. How lunar lava tubes can be used for human purposes?

A81. Lunar lava tubes could be an invaluable resource for supporting long term human presence on the moon. These tubes could provide shelter, human settlement, storage, and create exploration opportunities for astronauts in the future. Some of the uses of these tubes and advantages are given below:

(1) Lunar lava tubes are protected from micrometeorite rain, from cosmic rays, from solar radiations. An underground colony would escape the extreme temperatures on the moon's surface.

(2) The tubes can be used for storage and warehousing and in-vacuum manufacturing.

(3) The regolith Moondust is harmful for hardware and human lungs. Lunar lava tubes will provide a dust free environment.

(4) Lava tube sites will be much simpler and easier places for building construction. We have to provide only pressurization and maintain it.

(5) Some scientists have proposed to shift genetic material from earth to a vault on the moon, where it could be preserved for a very long time. Lunar lava tubes would be ideal places for this purpose where these will be protected from radiation and temperature variations. Genetic samples buried there could last billions of years without damage, allowing future humans to uncover the samples and replenish the earth after a catastrophe.

There are some challenges. The required pressurizing of lava tubes presents several hurdles such as sealing methods and sealant composition, pressurization stress, importation of astronomical volumes of nitrogen, etc. from earth.

References

1. *Kokh, Peter. Settling into a Lava Tube. Moon Miners' Manifesto, #100 November 1996, Section 6.9.3.2. 100.of the Artemis Data Book* http://adb.moonsociety.org/06/09/03/02/100/lava-tube-settling.htmlSettling into a Lava Tube
2. *Tomaswick, Andy. A Plan to Store Seeds, Spores, Eggs and Sperm on the Moon for Safekeeping. Universe Today. March 21, 2021.* https://www.universetoday.com/150630/a-plan-to-store-seeds-spores-eggs-and-sperm-on-the-moon-for-safekeeping/

Q82. Are lunar lava tubes intact, stable, and not prone to future collapse and asteroidal bombardment?

A82. Scientists feel that lava tubes that have survived to this day are likely to continue to do so for the rest of time. The vast bulk of major asteroidal bombardment to the moon took place in the first billion years of the moon's history. Lunar lava tubes, not subject to any sort of active geological forces or weathering are perhaps the safest, most stable, protected volumes to be found anywhere in the solar system.

Ridges and Cracks / Dorsa

Q83. What are wrinkle ridges on the moon?

A83. Wrinkle ridges have been seen on the surface of the moon for over a century. Studies of these began as early as 1885, with telescopic photographs, and continued beyond the Apollo era, with satellite and lander observations

A wrinkle ridge, also known as mare ridge, on the moon is a type of surface feature commonly found on lunar maria. Lunar wrinkle ridges are one of the most common linear to sinuous topographic highs on the lunar surface. They are also asymmetric, with one side of the ridge being higher than the other. These surface features are formed on the mare surface which can be hundreds of kilometres long, tens of kilometres wide, and hundreds of metres high. They are sometimes called veins as they look very much like the veins in the skin of a human body.

These wrinkled ridges suggest that the moon is shrinking. Using the 5-metre resolution stereo camera on India's Chandrayaan 2 orbiter and data from NASA LRO's laser altimeter, scientists have better mapped the structure of the prominent 220 kilometres long wrinkled ridge Dorsa Geikie. This gave them a better insight into the crustal stresses in the region and when the feature formed. Such studies are important to understand how the moon has shrunk over time.

Features like this have been found on several planets throughout the solar system, including the moon, Mars, Mercury, and Venus.

Q84. How are wrinkle ridges formed on the moon?

A84. We do not understand how wrinkle ridges are formed. Scientists in the late nineteenth century believed that wrinkle ridges are created when the basaltic lava cooled and contracted. Data collected during the Apollo mission suggested that wrinkle ridges were the surface expressions of thrust faults in the underlying crust. Now the thinking is that wrinkle ridges

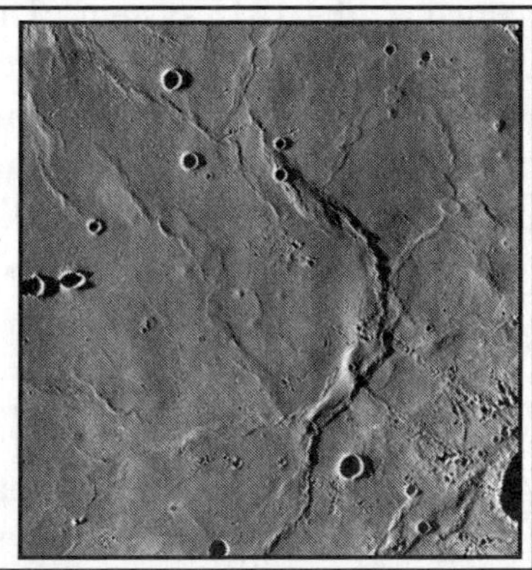

Bow of wrinkle ridges, called Dorsa Geikie, near the centre of Mare Fecunditatis on the moon.
Credits: NASA, Wikimedia Commons
Source: Part of mosaic of photos by NASA's Lunar Reconnaissance Orbiter (LRO)
Author: NASA,
Date: January 3, 2010
https://commons.wikimedia.org/wiki/File:Bow_of_wrinkle_ridges_in_central_Mare_Fecunditatis.png
Figure in the public domain

form by buckling of the mare areas and their surroundings. However, some scientists believe that recent findings from the Lunar Reconnaissance Orbiter Camera (LROC) may challenge this current understanding of wrinkle ridge formation. LROC images have identified some wrinkle ridges that are considerably different from the ones seen before and it remains to be seen if these new wrinkle ridges will again change our understanding of how these enigmatic features form. Scientists plan to target this area for further data acquisition, because only more data from LRO and further research will help solve the mysteries of the wrinkled moon.

Reference

Antonenko, Irene. A Wrinkled Moon. Universe Today. January 16, 2012
https://www.universetoday.com/92603/a-wrinkled-moon/

Q85. How are wrinkle ridges named?

A85. Wrinkle ridges are one of 18 different categories of lunar features recognized in the current system of International Astronomical Union (IAU) nomenclature. Wrinkle ridges are named with the Latin designation dorsum (plural dorsa). The term is used as prefix to the feature name. The term was introduced in 1973 to recognize wrinkle ridges that had not previously been named in the IAU system. The dimension given for a dorsum usually refers to the length of the ridge. The standard IAU nomenclature uses the names of people to identify wrinkle ridges on the moon. Except for the minor feature Dorsum Thera, the moon's ridges have been named exclusively after earth scientists.

Lunar swirl (Albedo)

Q86. What are lunar swirls?

A86. Lunar swirls are unusually bright and mysterious lunar features found across the moon's surface. They have a higher albedo than the surrounding surface. They have been identified on the lunar maria as well as highlands. They coincide with regions of the magnetic field of the moon. Every swirl has an associated magnetic anomaly, but not every magnetic anomaly has an identifiable swirl. Their origin has been debated since their identification in the 1960s. Scientists believe that swirls are formed by the deflection of the solar wind by intense local magnetic fields that shield the surface from the darkening effects of the solar wind (more on this later).

Reiner Gamma is one of the most visible lunar swirls from earth. It is associated with a localized magnetic field. This is one of the strongest localized magnetic anomalies on the moon.

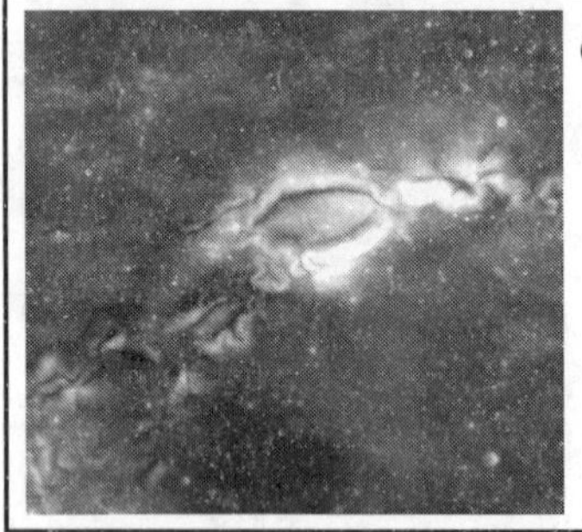

Colour image of the Reiner Gamma albedo feature on the moon.
Credit: Wikimedia Commons
Author: NASA (Created using the USGS PDS Map-a-planet website)
Date: July 7, 2006
https://commons.wikimedia.org/wiki/File:Reiner_gamma.jpg
File is in the public domain

Q87. How are lunar swirls formed?

A87. The formation of lunar swirls is a mystery. There are three leading theories/models for swirl formation:

1. Cometary impact model

According to this theory, the high albedo of the swirls is the result of an impact with a comet. The impact would cause scouring of the top-most surface regolith by the coma's turbulent flow of gas and dust, which exposed fresh material and redeposited the fine, scoured material in discrete deposits.

2. Solar wind shielding model

This theory suggests that the swirls are formed because the lighter-coloured regolith is protected from the solar wind, due to a magnetic anomaly. The swirls represent exposed silicate materials whose albedos have been selectively preserved over time. This model suggests that swirl formation is a continuing process, which began after the creation of the magnetic anomaly. Mathematical simulations conducted in 2018 showed that lava tubes could have become magnetic as they cooled, which would provide a magnetic field consistent with the observations near the lunar swirls. Although the moon does not have a global magnetic field, localized magnetic fields have been observed in swirl regions.

3. Dust transport model

The dust transport model proposes that weak electric fields created by interaction between the crustal magnetic anomalies and the solar wind plasma could attract or repel electrically charged fine dust. High albedo, feldspathic material is the dominant component of the finest particles of lunar soil. Electrostatic movement of dust lofted above the surface during terminator crossings could cause this material to preferentially accumulate and form the bright, looping swirl patterns.

Q88. Which theory of lunar swirls formation is supported by recent measurements/observations?

A88. Out of the above theories, Satellite measurements have supported the solar wind shielding model. Observations by the Lunar Reconnaissance Orbiter (LRO) support that solar wind is being deflected by a magnetic field. Spectral observations by the moon Mineralogy Mapper instrument on board Chandrayaan-1 confirmed that the lighter-coloured regions are deficient in hydroxide, which also supports the hypothesis that solar wind is being deflected in the pale areas.

Research using data from NASA's ARTEMIS (Acceleration, Reconnection, Turbulence and Electrodynamics of the moon's Interaction with the sun) mission explain how the solar wind and the moon's crustal magnetic fields work together to give the moon a distinctive pattern of darker and lighter swirls. The sun releases a continuous outflow of the solar winds (particles and radiation) which reach to the planets, moons, and other bodies in our solar system. We are largely protected from the damaging effects of the solar wind because earth's magnetic field deflects most of the solar wind particles. But the moon does not have a global magnetic field, it has only localized spots which act as a magnetic "sunscreen" and deflect solar wind particles to the areas just around where chemical reactions with the regolith darken the surface. This creates the distinctive swirls of darker and lighter material that are so prominent they can be seen from the earth.

Scientists are planning future experiments to further study these swirls. As of 2018, a CubeSat mission concept is under study at NASA, with the goal of understanding the formation of the lunar swirls. The proposed BOLAS (Bi-sat Observations of the Lunar Atmosphere above Swirls) is a spacecraft mission concept that would orbit the moon at very low altitude to study the lunar surface. The concept involves two small identical CubeSat satellites connected vertically above the lunar surface by a 25 km (16 miles) long tether (a space tether is a long strand of fibers used to couple a space craft with another mass which allows for an exchange of momentum or energy. Tethers can be used to change the orbits of satellites without consuming any propellant, or to transport payloads from a planet to low altitude orbit). The mission goal would be to understand the hydrogen cycle on the moon, dust weathering, and the formation of lunar swirls.

NASA is planning a mission in 2024 to the lunar swirl of Reiner Gamma. The mission will have a joint payload suite called

Lunar Vertex on a lander and rover to map the magnetic field from the swirl's surface and study its properties in-situ. This will be the world's first mission to a lunar swirl and will provide insights into how they form and evolve, as well as better our understanding of the moon's interior and core, and its magnetic evolution.

References

1. *Lunar swirls. Wikipedia. Last edited on March 18, 2021,* https://en.wikipedia.org/wiki/Lunar_swirls
2. *Phillips, Tony Dr. Lunar Swirls. NASA Science. June 26, 2006* https://science.nasa.gov/science-news/science-at-nasa/2006/26jun_lunarswirls/
3. *Understanding Lunar Swirls. NASA Press Release. SpaceRef (Moon Today). February 28, 2019* http://spaceref.com/moon/understanding-lunar-swirls.html

Moon Mapping

Q89. How has the moon has been mapped through time?

A89. The historical development of mapping of the moon is quite an interesting tale. As early as 450 BC, a Greek philosopher, Democritus believed that there were lofty mountains and hollow valleys on the moon's surface. In the absence of photographic techniques, early astronomers used to sketch the observed details of the moon's surface. Some of them show sharp surface features. The Robotic mission later demonstrated that this is not the case. The surface features are much more rounded due to a long history of impacts. As soon as telescopes provided the moon's closer look, people started making the moon's maps. There have been continuous improvements in detailing. Serious work began by the end of the fifteenth century. Some of the important chronological details regarding moon-mapping are summarized below:

1603: Based on naked-eye observation, William Gilbert, an English physician, physicist, and natural philosopher made the first lunar drawings.

1609: Thomas Harriot, an English astronomer and map maker, produced a series of exquisite lunar drawings, one of which is dated July 26, 1609. This is generally considered to be the first astronomical use of the telescope.

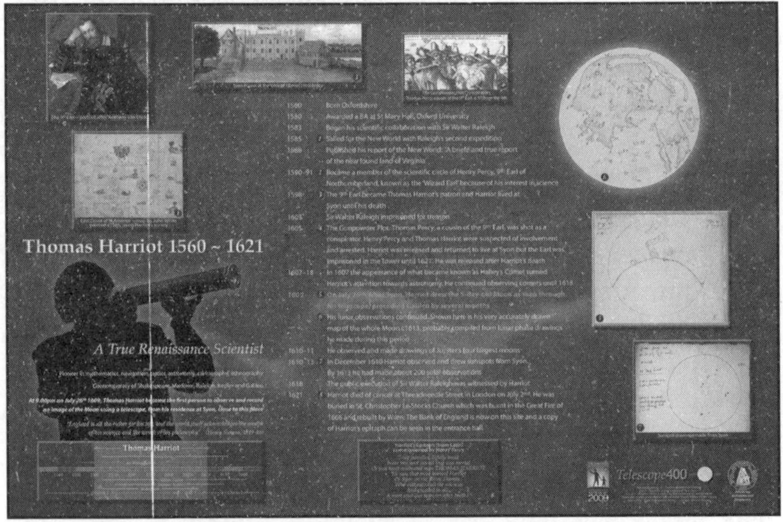

This plaque was erected in July 2009 in the grounds of Syon House (West London) to commemorate the 400th anniversary of Thomas Harriot's drawings of the moon using a telescope. This is generally considered to be the first astronomical use of the Telescope.
Credits: Wikimedia Commons, Source/Author: The group organising the Telescope 400 event (July 2009)
Date: September 7, 2011
Figure in Public Domain
https://commons.wikimedia.org/wiki/File:Thomas_Harriot_Memorial_Plaque_(Syon_Park,_London).jpg

1610: Galileo produced a set of six watercolour sketches of the moon in its various phases as he observed the moon through a telescope in November-December of 1609. He published a short astronomical treatise (Sidereus Nuncius) in New Latin on March 13, 1610. It was the first published scientific work based on observations made through a telescope, and it contains the results of Galileo's early observations of the imperfect and mountainous moon, the hundreds of stars that were unable to be seen in either the Milky Way or certain constellations with the naked eye, and the Medicean Stars (later Galilean moons) that appeared to be circling Jupiter.

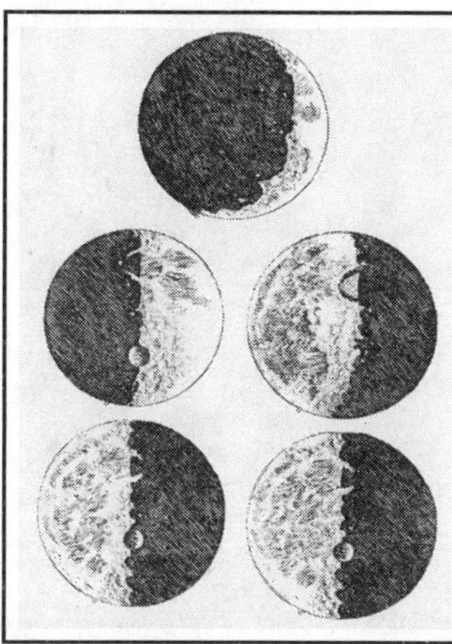

Galileo's sketches of the moon published in March 1610.
Credits: Wikimedia Commons
Author: Galileo
https://commons.wikimedia.org/wiki/File:Galileo%27s_sketches_of_the_moon.png
Figure in Public Domain

1645: Michael Florent van Langren, a Dutch astronomer made a map of the moon using a telescope. It probably was the moon's first map. This was a detailed map with nomenclature of the moon's surface details. The map had over 300 names of the moon's surface features: land masses, seas, craters, and peaks. Some 50 names, introduced by him are still used. This map enabled mariners to find their positions just by observing which of the moon's craters were illuminated by the sun and to what extent.

1647: Polish astronomer Johannes Hevelius made the First Atlas of the moon with detailed maps of the moon. Hevelius devoted four years to charting the lunar surface and published his results in Selenographia, sive Lunae description (Selenography, or A Description of the moon), a work which entitles him to be called "the founder of lunar topography". Hevelius documented every crater, slope, and valley that he could see with his telescope.

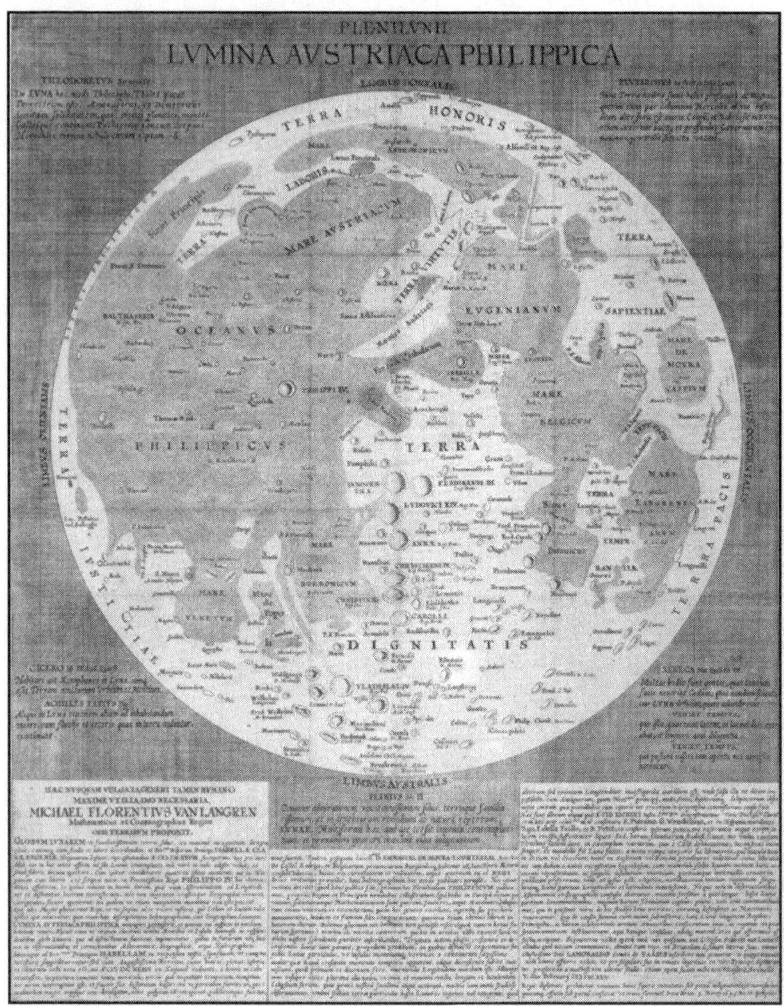

Map of the Moon, published by Michael van Langren in 1645 in his book Plenilunii lumina Austriaca Philippica
Credits: Wikimedia Commons, Author Michael van Langren, 1645
https://commons.wikimedia.org/wiki/File:Langrenus_map_of_the_Moon_1645.jpg
Figure in Public Domain

He introduced his own naming system based on geographical names. His atlas served as the standard reference work for over a century.

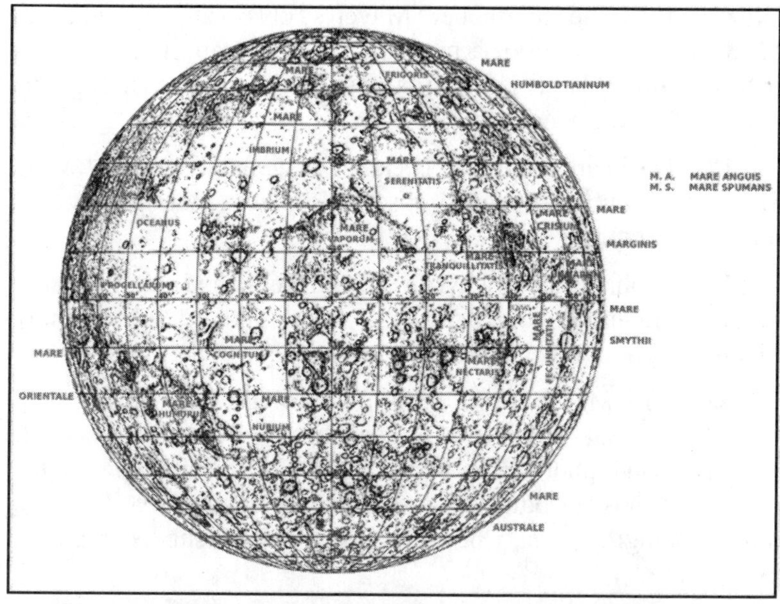

Map of the Moon from Selenographia, (sive Luna description 1647)
Credits: Wikimedia Commons, Author Johannes Hevelius (1611–1687), Date 1647
Source http://www.e-rara.ch/zut/content/pageview/160524
https://commons.wikimedia.org/wiki/File:Hevelius_Map_of_the_moon_1647.jpg
Figure in Public Domain

1651: Two Jesuit scholars of Italy, Giovanni Battista Riccioli and Francesco Grimaldi devised the modern nomenclature in 1651. Many of the craters were named after ancient Greek personalities (such as Plato, Atlas, and Archimedes), Ancient Roman empires (such as Julius Caesar, Tacitus, and Taruntius), scholars, writers, and philosophers from medieval Europe, and saints of the Catholic church. Later maps dropped the "St." from names.

1750: Johann Meyer produced the first reliable set of lunar coordinates that permitted astronomers to locate lunar features.

1775: Tobias Mayer (1723-1762), a mathematician, astronomer, and German cartographer, undertook lunar cartography with a new precision by using a micrometer integrated in the telescope. Mayer's observations began in 1748 for the creation of a map and a world map. The map was published in 1775. It is the first map of the moon that contains longitudes and latitudes.

1779: Lunar mapping became systematic when Johann Schröter began meticulous observation and measurement of lunar topography.

1834: Johann Heinrich von Mädler published the first large map of the moon, comprising 4 sheets and he subsequently published The Universal Selenography.

1840: In March 1840, John William Draper, an English-born American scientist, philosopher, physician, chemist, historian, and photographer produced a daguerreotype of the moon and thus introduced photography to astronomy. By 1890 lunar photography had become a recognized subdiscipline of astronomy.

> Note: daguerreotype: Named after the inventor, Louis Jacques Mandé Daguerre, daguerreotype is a photograph taken by an early photographic process employing an iodine-sensitized silvered plate and mercury vapor.

The twentieth century brought more advances.

1919: The procedure of naming the moon's features has been assumed by the International Astronomical Union (IAU) in 1919.

1967: NASA's Lunar Orbiter missions completed photographic mapping of the moon.

1969: Famous amateur astronomer, Sir Patrick Moore, produced a hand drawn map of the moon from his own observations using his homemade telescope at his home in Selsey, Sussex. These detailed pen and ink maps of the moon's surface were used by NASA as part of their preparations for the moon landing. In 2007, Ordnance Survey (OS) printed 250 signed limited editions of the maps.

1994: NASA's Clementine mission conducted multispectral mapping of the moon.

2011: NASA's Gravity Recovery and Interior Laboratory (GRAIL) mission mapped the moon's gravitational field in detail.

2020: For the first time, the entire lunar surface has been completely mapped and uniformly classified by scientists from the USGS Astrogeology Science Center, in collaboration with NASA and the Lunar Planetary Institute. The lunar map is known as the Unified Geologic Map of the moon. Scientists used information from six Apollo-era regional maps along with updated information from recent satellite missions to the moon. The Terrain Camera on the recent SELENE (Selenological and Engineering Explorer) mission led by the Japan Aerospace Exploration Agency (JAXA) provided data for the moon's equatorial region, while data for the north and south poles was supplemented with NASA's Lunar Orbiter Laser Altimeter data.

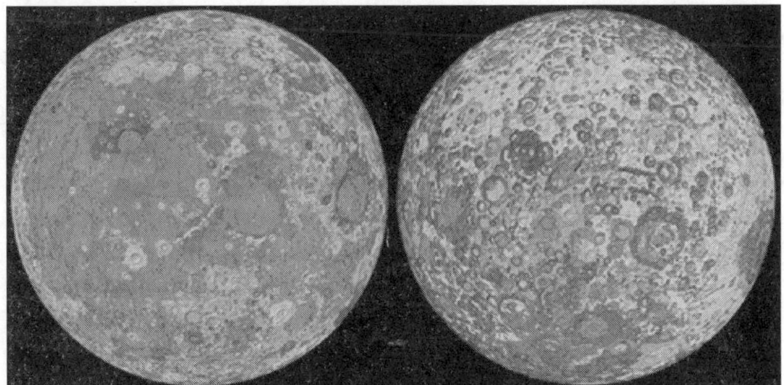

Unified Geologic Map of the Moon showing the geology of the moon's near side (left) and far side (right). This geologic map is a synthesis of six Apollo-era regional geologic maps updated using data from recent satellite missions.
Credit: NASA/GSFC/USGS.
https://www.usgs.gov/news/usgs-releases-first-ever-comprehensive-geologic-map-moon
Figure in Public Domain

❑

CHAPTER 5

Lunar Atmosphere

Q90. Is there an atmosphere on the Moon?

A90. A quick answer to this question is “No”. However, a better answer is "sort of." For most practical purposes, the moon is surrounded by a vacuum. Until recently, most everyone accepted this conventional wisdom. However, according to NASA's Lunar Atmosphere and Dust Environment Explorer website, recent studies confirm that the moon does indeed have an atmosphere consisting of some unusual gases, including sodium and potassium, which are not found in the atmospheres of earth, Mars, or Venus. But it is an extremely thin atmosphere.

Q91. Why doesn't the moon have an atmosphere?

A91. There are three reasons why the moon has no atmosphere:

(1) The primary reason is the weak gravity of the moon. Strong gravity is needed to hold an atmosphere into place. Gases are lost to outer space with weak gravity.

(2) The moon is currently not very active geologically (almost dead). There are currently no significant processes that allow for abundant gases to supply an atmosphere. The early moon may have had an atmosphere when it was geologically active, but the crust of the moon is currently “frozen” into place and

thus there are very little gases being emitted as compared to earth.

(3) The moon is relatively close to the sun. The solar wind from the sun helps strip away a weak atmosphere.

Q92. What is the moon's thin atmosphere made of and what are the sources?

A92. The Apollo 17 mission deployed an instrument called the Lunar Atmospheric Composition Experiment (LACE) on the moon's surface. It detected small amounts of helium, argon, and possibly neon, ammonia, methane, and carbon dioxide. Still, we only have a partial list of what makes up the lunar thin atmosphere. Many other species are expected. Whatever little we have in the moon's atmosphere, there are several possible sources. These include high energy photons and solar wind particles knocking atoms from the lunar surface, chemical reactions between the solar wind and lunar surface material, evaporation of surface material, material released from the impacts of comets and meteoroids, and out-gassing from the moon's interior. But which of these sources and processes are important on the moon is still not known?

Q93. What are the differences between the atmospheres of earth and the moon?

A93. The moon's very thin atmosphere has an infinitesimal amount of air when compared to earth's atmosphere. Compare this: At sea level on earth each cubic centimeter contains 10 ^19 (10,000,000,000,000,000,000) molecules. The lunar atmosphere has less than 10^6 (1,000,000) molecules in the same volume. The pressure of this small mass is around 3×10^−15 (0.000000000000003) atm, and the total mass is less than 10 metric tons. It is what we consider to be a very good vacuum on earth. In fact, the density of the atmosphere at the moon's surface is comparable to the density of the outermost fringes of earth's atmosphere where the International Space Station orbits. (Therefore, we normally say that the moon has no atmosphere).

One of the critical differences between the atmospheres of earth and the moon is how atmospheric molecules move. At the surface of earth, the molecules' motion is dominated by collisions between the molecules. However, the moon's atmosphere is so thin, atoms and molecules almost never collide. Instead, they are free to follow arcing paths. This type of thin, collision-free atmosphere, that extends all the way down to the ground is known as "surface boundary exosphere." Scientists believe this may be the most common type of atmosphere in the solar system.

Q94. Did the ancient moon once have an appreciable atmosphere?

A94. Scientists at NASA's Marshall Space Flight Centre and the Lunar and Planetary Institute (LPI) in Houston found that the moon likely had an ancient, temporary atmosphere around 3 to 4 billion years ago. The atmosphere was caused by ancient volcanos releasing gases into the lunar sky faster than they could escape to space. The atmosphere, which was about 2 per cent the density of earth's but twice that of Mars, lasted for around 70 million years. The study found that the moon's magmas carried gas components, such as carbon monoxide, the ingredients for water, sulfur, and other volatile species. The ancient lunar atmosphere was likely capable of weathering rocks and producing windstorms. Perhaps most importantly, it could be a source for some, if not all, of the water detected on the moon.

The ancient lunar atmosphere was eventually stripped away by solar winds and dissipated into space.

Reference

John, Tara. NASA: The Moon Once Had an Atmosphere That Faded Away. Time. October 9, 2017 https://time.com/4974580/nasa-moon-had-atmosphere-volcanoes/

Q95. What are the consequences of the moon's non-existent or thin atmosphere?

A95. The moon's non-existent or thin atmosphere has several consequences, such as:

(1). Black sky, brighter stars: The moon's sky is black instead of blue. The earth's sky is blue due to Raleigh scattering. Gases and particles in the earth's atmosphere scatter sunlight in all directions. Blue light is scattered more than other colours because it travels as shorter, smaller waves and the smallest wavelengths of light tend to scatter easier. Since there is no atmosphere on the moon, we see a black sky. A pitch-black background makes stars appear brighter and sharper in the moon's sky but practically no atmosphere means no twinkling of the stars.

(2). No sound: Sound waves need a medium to travel. In the absence of any atmosphere (medium), a deadly silence prevails on the moon's surface. So, the moon is a silent world.

(3) No weather: There simply is no weather on the moon. Weather requires the presence of an atmosphere, like the blanket of gas molecules that surrounds earth.

(4). Things change very slowly on the Moon's surface: The moon's surface has remained much the same for over 3 billion years. Apollo 15 astronauts found a rock which is estimated to be 4.5 billion years old. The rock is as old as the moon. Such old rocks are not found on earth. Rain, wind, weathering, earthquakes, volcanism, change everything on earth over a period. Nothing of that sort happens on the moon's surface. The footprints left behind by Apollo astronauts will remain there till they get wiped out by some fresh impact by comets or, meteors.

(5). Temperature variance: The air that surrounds our earth acts as a blanket to keep us warm and comfy. But the moon, since it does not have such a blanket, gets much colder and hotter than the earth. When sunlight hits the moon's surface, the temperature can reach 260 degrees Fahrenheit (127 degrees Celsius) and when the sun goes down, temperatures can dip to minus 280 F (minus 173 C). Shadowed regions also drop to such low temperatures. The poles can get even colder. Orbiting spacecraft instruments have measured the surface temperatures of moon's polar regions and found these dark areas to be extremely cold, reaching surface temperatures as low as minus 382 degrees F (minus 230 degrees Celsius). That makes the polar regions among the coldest parts of the solar system. The south pole is among the coldest parts of the solar system and may be colder than what we expect from places like Pluto.

(6). **Shadows on the Moon's surface are darker than on the earth, but nights are not as dark:** On the earth, air molecules scatter light (Rayleigh scattering) and objects are well-lit even in the shade. This is due to this scattering effect that we can read a magazine under an umbrella at the beach. The shadows on earth are also not very dark due to this scattering. As the moon has no atmosphere, there is no Rayleigh scattering, and the shadows are pitch dark. However, even at night, it is never too dark on the moon's near side. A bright sunlit disk of the earth, over 40 times brighter than full moon, never allows it to be pitch dark. You can safely drive around the big boulders on the moon's surface in the glow of earthshine. If you visit the moon's far side however, you will neither see the earth nor its shine

Q96. Does the moon have weather?

A96. Our moon does not have weather like we experience on earth. Our earth has weather because of the earth's thick troposphere, which extends up to 6-7 miles (~10 km). About 75-80 per cent of the mass of our atmosphere is within the troposphere and most of the moisture too. But since the moon does not have an appreciable atmosphere, there is no earth like weather. However, discoveries of water ice on the moon suggest that there could be some type of "water cycle."

But we do have extreme temperatures on the moon ranging from boiling hot to freezing cold depending on where the sun is shining. The moon rotates on its axis in about 27 days. Daytime on one side of the moon lasts about 13 and a half days, followed by 13 and a half nights of darkness. When sunlight hits the moon's surface, the temperature can reach 260 degrees Fahrenheit (127 degrees Celsius). When the sun goes down, temperatures can dip to minus 280 F (minus 173 C).

There are also no seasons on the moon. We have seasons on earth because of the earth tilts on its axis at roughly 23.5 degrees. Because of the axial tilt, as the earth orbits the sun, different parts of the earth are heated differently. For this reason, when it is summer in the northern hemisphere, it is winter in the southern hemisphere. The moon has only a 1.54-degree tilt. So, there is virtually no seasonal variability. But, the poles have small regions where the sun doesn't rise for around six months. It was there that the lowest temperatures were measured.

Reference

Shepherd, Marshall. Does Our Moon Have Weather? Forbs – Science. July 18, 2019
https://www.forbes.com/sites/marshallshepherd/2019/07/18/does-our-moon-have-weather/?sh=4e0e4bbb45b2

Q97. Why is there so much temperature variations on the moon?

A97. The earth and the moon receive similar amounts of energy from the sun. In the case of the earth, the atmosphere deflects and absorbs some of that heat. As the sun's rays strike the gas molecules surrounding the earth, those molecules absorb some of the energy and pass it on throughout the atmosphere, warming the entire earth rather than only direct sunlit areas. Such diffusion of energy reduces the maximum temperature. Since the moon has no such protective blanket, its maximum temperatures are scorching. There is also the greenhouse effect. An atmosphere traps the sun's energy. When the sun's energy makes it through the earth's atmosphere and strikes the surface, it reflects off the surface and back toward space. In much the same way that gas molecules absorbed and trapped energy on its way in, these molecules trap and reflect energy on its way out, keeping the earth warm. On the moon, however, any energy that reflects off the surface simply dissipates into the vacuum, which is why the shadowed areas of the surface become extremely cold.

Q98. How do astronauts survive huge temperature variation on the moon?

A98. 1. All trips to the moon were planned for lunar dawn, to ensure the moon's surface hadn't had time to heat up fully to its daytime temperature.

2. Out of the three ways of heat transfer, conduction, convection, and radiation, only two are possible: conduction and radiation. Convection is not possible because it requires a fluid medium, and the moon has no atmosphere. Conduction is from direct contact of astronauts' feet with the moon's surface. But, since the moon's soil (regolith) doesn't conduct heat well and the astronauts' boots were insulated, conduction is minimal. Radiation is both directly from the sun and from the sun's reflection. The astronauts' spacesuits were designed to reflect almost 90 per cent of the light that reaches it, so very little heat would have transferred to the astronauts. This shows that even though huge temperature variations occur on the moon's surface, astronauts were never actually exposed to them.

❑

CHAPTER 6

Lunar Gravity

Q99. Does the moon have gravity?

A99. Everything that has mass has gravitational force associated with it. Therefore, the moon does have gravity.

We use the term "surface gravity" in reference to the downward "pull" that objects experience when resting or moving on a larger body. A body's surface gravity is proportional to its mass, but inversely proportional to the square of its radius. Earth's average surface gravity is about 9.8 metres per second per second (9.8 m/s^2). This means that when an object is tossed off a building top it accelerates toward the ground at 9.8 metres per second per second. The moon's surface gravity is about 1/6th as powerful or about 1.6 m/s^2. To compare, sun's surface gravity is 274 m/s^2. The objects fall slowly on the moon because their acceleration toward the surface is only 1/6th what it would be on earth.

Earth's gravitational force keeps us on its surface. When we see astronauts on the moon, they bounce around (or can jump higher than we can on earth) because gravity there is less strong. When on earth, a fully suited Apollo astronaut weighed about 500 pounds, equipment included. His weight was only about 80 pounds on the moon.

Q100. Why is the moon's gravity so uneven?

A100. Over the years, scientists have observed that the moon's gravity is stronger in some regions than others, creating a "lumpy" gravitational field. This first became apparent in 1966 when the Soviets noticed that their moon orbiter Luna 10 was deviating from its orbit significantly. America's Lunar Orbiter missions also faced similar situation. The cause of the lumpy gravity is an uneven mass distribution below the lunar surface. The concentrated mass areas are known as mass concentrations, or "mascons." Exactly how these mascons came into being remained a mystery till now.

This image shows the variations in the lunar gravity field as measured by NASA's Gravity Recovery and Interior Laboratory (GRAIL) during the primary mapping mission from March to May 2012. Red represents mass excesses, and blue represents mass deficiencies. Credit: Wikimedia Commons, NASA/ JPL-Caltech/MIT/GSFC https://commons.wikimedia.org/w/ index.php?curid=23051106 Figure is in Public Domain

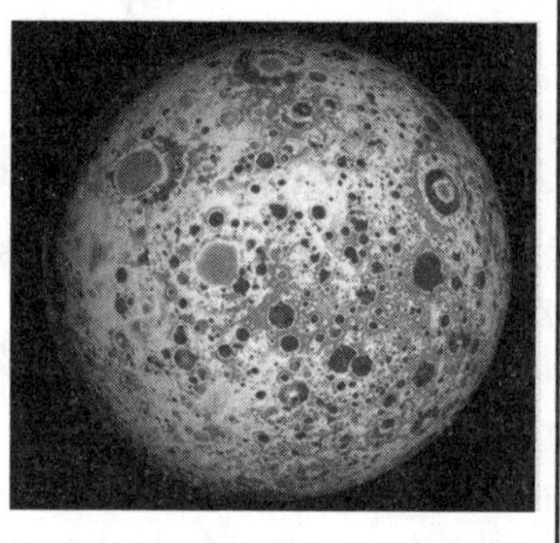

NASA's Gravity Recovery and Interior Laboratory (GRAIL) mission data reveal that mascons gravitational fields resemble a bull's-eye pattern: a centre of strong gravity surrounded by alternating rings of weak and strong gravity. Scientists now think they have an explanation for this. When an asteroid crashes into the moon, it sends material flying out, creating a dense band of debris around the crater's perimeter. The impact sends a shockwave through the moon's interior, reverberating within the crust and producing a counterwave that draws dense material from the lunar mantle towards the surface, creating a dense centre within the crater. After hundreds of millions of years, the surface cools and relaxes, creating a bull's-eye that matches today's gravitational pattern.

Mare Orientale gravity map showing a bull's eye pattern. Red corresponds to mass excess and blue to mass deficiencies
Credits: Wikipedia, NASA/JPL-Caltech/Arizona State University; generated by James Stuby, Source JMARS, 2016
https://en.wikipedia.org/wiki/File:Orientale_basin_GRAIL_gravity.jpg
Figure is in the public domain

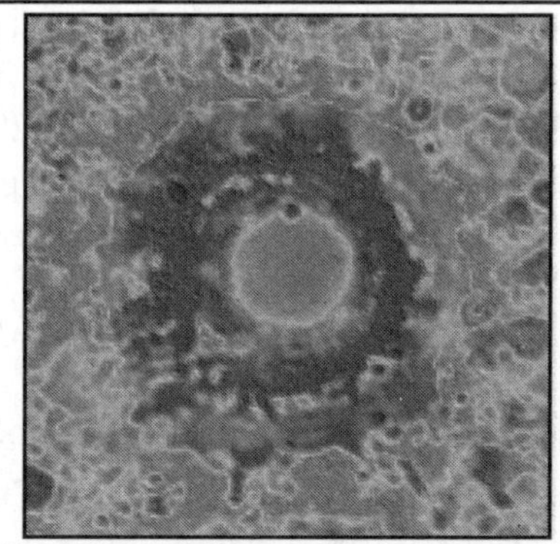

It is interesting to note that a mysterious large mass of material (a minimum mass of 2.18 × 10^18 kg and likely extends to depths of more than 300 km (190 mi) has been discovered beneath the moon's South Pole-Aitken basin, the largest crater in our solar system (roughly 2,500 km (1,600 mi) in diameter and between 6.2 and 8.2 km (3.9–5.1 mi) deep). According to Peter B. James and others, it may contain metal from an asteroid that crashed into the moon and formed the crater. One of the explanations of this extra mass is that the metal from the asteroid that formed this crater is still embedded in the moon's mantle. A sufficiently dispersed core of the asteroid that made the impact could remain suspended in the moon's mantle until the present day, rather than sinking to the moon's core. Another possibility is that the large mass might be a concentration of dense oxides associated with the last stage of lunar magma ocean solidification.

Note: Mascons have also been found on Mercury and Mars. There is even a mascon on earth around Hawaii.

References

1. *James, Peter B. Smith, David et al. Deep Structure of the Lunar South Pole-Aitken Basin. AGU Geophysical Research Letters. Volume 46, Issue 10. Pages: 5009-5631. May 28, 2019 (First published: April 5, 2019).*

https://agupubs.onlinelibrary.wiley.com/doi/full/10.1029/2019GL082252
https://doi.org/10.1029/2019GL082252

2. *Mehta, Jatan. How we got to know the Moon's gravity is lumpy. Planetary Science, Space Missions. June 19, 2020.*
https://blog.jatan.space/p/the-moons-lumpy-gravity-field

Q101. Is low gravity of the moon a health hazard?

A101. This is one of the areas where medical experts are still trying to fully understand. Based on a series of space station experiments dating back to the 1970s and earthbound studies, scientists have developed a growing understanding of how zero gravity affects the human body as well as learning how astronauts can most effectively live and work in free fall. They believe that the level of gravity on the moon could be harmful on bones, muscles, and other organs. Moon's one-sixth gravity does not put enough weight on our bodies to protect against loss of bone mass, muscle strength, and heart pumping capacity. Experts say that effective exercise regimes can be adapted for life on the moon to prevent ongoing atrophy. However, based on experience of the International Space Station (ISS), some experts believe that long duration visitors will still experience some of low gravity's deleterious effects.

> Note: Other than low gravity, astronauts will face two other sources of danger on the moon: radiation and moon dust (regolith). More on this later

Q102. How does the gravitational force of the moon affect earth?

A102. The Moon's gravitation keeps earth's conditions stable. Without the moon's gravitational force, life would be impossible on earth. The water that remains on the surface, the stable climate, and the stable rotational speed of the earth are all results of the moon's gravitational force.

Weather

The Moon's gravitational force produces ocean tides. A world without tides would have very different weather systems. Tides produce ocean currents moving warm or cool water about the earth. Warm ocean currents bring warmer wetter weather, while cold ocean currents bring colder drier weather. The moon's gravitational pull influences tidal currents and so the movement and mixing of the upper layers of the ocean water.

Ice, land, and air

The moon generates tidal currents and waves at the surface as well as deeper in the ocean. These currents and waves may either melt or break up sea ice.

Warth's land tides are like ocean tides. The land deforms and bulges just as the sea does and is thought to trigger volcanic activity and earthquakes.

The moon's gravitational forces also cause oscillations in the earth's atmosphere resulting low and high atmospheric pressures. Lower pressure is known to result in cold wet weather and higher pressure in calmer more pleasant weather.

Reference

Latham, Katherine. The subtle influence of the moon on earth's weather. BBC Future. August 23, 2021 https://www.bbc.com/future/article/20210820-the-subtle-influence-of-the-moon-on-earths-weather

❑

CHAPTER 7

Magnetic Field of the Moon

Q103. Does the moon have a magnetic field?

A103. The magnetic field of the moon is currently very weak, less than one-thousandth as powerful as earth's. The major difference is the moon does not have a dipolar magnetic field now as would be generated by a geodynamo in its core. So that the magnetization present is varied, and its origin is almost entirely crustal in location. But when the rocks brought back by the Apollo astronauts were analyzed the researchers were surprised to find residual traces of magnetism. In other words, there are localized regions on the moon's surface up to several hundred kilometres in size where magnetic field prevails.

Using data from magnetometers onboard NASA's Lunar Prospector and JAXA's SELENE orbiters, researchers have made a large-scale map of local, anomalous magnetic fields on the moon with higher resolution than before.

Note: Geodynamo is the mechanism by which the earth's magnetic field is generated, believed to resemble a self-exciting dynamo and to which involve circulatory and convective motions and electrical currents within the earth's liquid outer core. earth's magnetic field is almost the same shape as the magnetic field around a bar magnet, so it is called a "dipolar magnetic field" (two poles).

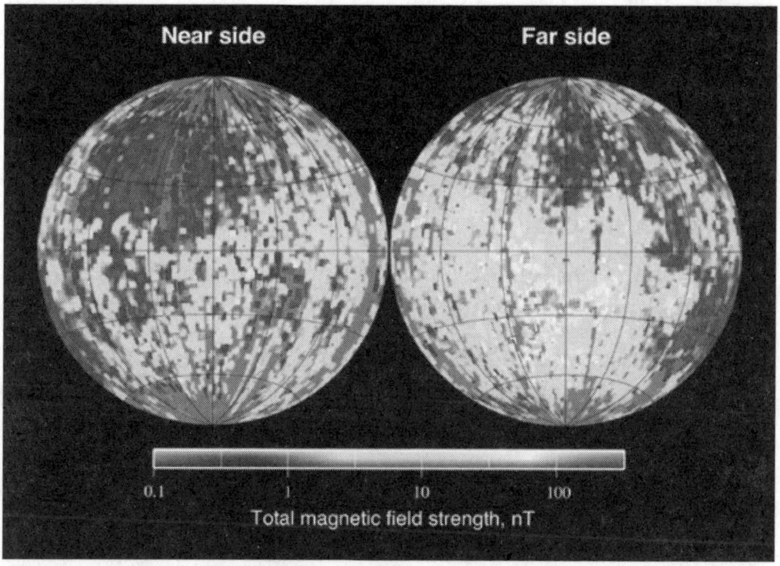

Total magnetic field strength at the surface of the moon as derived from the Lunar Prospector electron reflectometer experiment.
Credit: Wikimedia Commons.
Author Mark A. Wieczorek
Date: created November 16, 2006.
Permission: As long as the author is attributed, feel free to use this as you please.
https://commons.wikimedia.org/wiki/File:moon_ER_magnetic_field.jpg

Q104. Why does the earth have a magnetic field, whereas the moon does not?

A104. A magnetic field is generated by a dynamo mechanism which is caused by the fluid motion of a conducting material (such as liquid iron). This process of a core dynamo is the principal explanation for global magnetic fields in planets throughout the solar system. The large magnetic field of earth is generated in its very large, rapidly rotating, molten nickel-iron outer core caused by the convection currents. But, compared to earth, the moon is a small body (1 per cent mass), rotates very slowly (almost 30 times slower), hence does not have a hot, viscous outer core. So, convection currents do not occur. Thus, it is not surprising that the moon has no global field.

Q105. Is varying magnetic field strengths of the moon related to its geology?

A105. Based on data from different lunar missions, Staci L. Tiedeken (currently in NASA) has reported the following:

1. Older materials of the moon tend to be more magnetic than younger materials.
2. The mare regions of the moon are less magnetic than the plains and terra regions.
3. No simple relationship was found between lunar magnetism and crustal thickness which is inconsistent with the hypothesis that the thicker crust is more magnetic since there is additional material.
4. There is not a consistent pattern of magnetic field enhancements near specific craters, but evidence for these strong magnetic signatures was present for Mare Crisium and Mare Nectaris. These results may support the hypothesis that ejecta materials are carriers of magnetism, and this may be the reason for their tendency to have higher magnetic field strengths.

Reference

Tiedeken, Staci L. "Magnetism and Geology of the Moon." MS (Master of Science) thesis, University of Iowa, 2017 https://doi.org/10.17077/etd.0vwyfonj

Q106. Was there a strong magnetic field on the moon earlier?

A106. The moon has no global magnetic field now. But based on study of the Apollo samples, scientists believe that there was a strong magnetic field on the moon over 3 billion years ago, as strong as earth's field today. Based on such studies, and measurements by instruments carried by the Apollo astronauts and from orbit by robotic missions, scientists feel that the internal mechanism powering the moon's global magnetic field changed dramatically between 3.6 to 3.1 billion years ago and weakened significantly towards the end of that period. It appears that a lunar core dynamo somehow started sometime in the first few hundred million years of the moon's existence, operated for a while, and then quit around 3.5 billion years ago. Scientists are researching to understand reasons why a magnetic field should start, then abruptly stop operating.

Q107. What is the mystery of the moon's Lost Magnetism?

A107. Moon's lost magnetism is one of the questions which remains unanswered. Now, two teams working separately seem to have found two complementary explanations.

(1) According to NASA's virtual institute SSERVI, Christina Dwyer and her team of the University of California, Santa Cruz suggest that the moon's core and its mantle rotate around slightly different axes. The boundary between them is not quite spherical, so their relative motion causes the fluid to mix around. Their model suggests that the angle between the mantle and the core has narrowed over time, while the distance between the moon and the earth has widened, causing the tidal forces to steadily decrease. While these forces used to be enough to generate a dynamo inside the moon, they are not anymore. Scientists have suggested that further research is needed to model complex physics of dynamos to investigate whether this could explain what happened on the moon.

(2). Michael Le Bars of the French National Centre for Scientific Research and the Universite Aix-Marseille in France with his colleagues suggest violent impacts as another explanation for a lunar magnetic field. They believe that the lunar mantle stirred the core was caused by impacts with large bodies which slammed into the moon and changed its rotation rate, thus creating a discrepancy between its mantle and its core.

Some scientists have expressed their opinion by suggesting that while both theories have a good chance of being correct, a third option, a combination of the two, is also possible. Future experiments on samples from very old lunar rocks will be required for testing their theories.

References

1. *Soderman/NLSI Staff. Mystery of Moon's Lost Magnetism Explained. sservi.nasa.gov*
https://sservi.nasa.gov/articles/mystery-moons-lost-magnetism-explained/
2. *Andrei, Mihai. Mystery of moon's past magnetism might be solved.*
ZME Science. In Astrophysics, Physics. November 10, 2011
https://www.zmescience.com/science/physics/moon-magnetism-10112011/
3. *Williams, Matt. Did the Moon Ever Have a Magnetosphere? Universe Today. August 16, 2021*
https://www.universetoday.com/152152/did-the-moon-ever-have-a-magnetosphere/

Q108. Is it a confirmed conclusion that the moon once had a Magnetosphere?

A108. No, few scientists still pose a question - did the moon Ever Have a Magnetosphere? NASA is planning to investigate this mystery through their ambitious Artemis mission.

The hypothesis that the moon once had a strong magnetic field which faded away quickly is based on studies of samples of lunar rocks returned by the Apollo astronauts. Scientists, however, could never explain how a small moon's core could create a strong magnetic field.

An international team of geophysicists led by John A. Tarduno of the University of Rochester, has recently studied the Apollo samples using a new scientific technique to re-examine the evidence for lunar magnetization. Their new technique allowed them to obtain more accurate readings of their magnetization without altering them, which may have been a factor in the past and led to misleading results. These samples didn't show any magnetization indicating that the moon has never possessed a long-lived magnetic field. The researchers determined that the earlier readings could be the result of impacts from meteorites or comets, not a magnetic field. When meteorites strike the moon, they create a dusty plasma – a gas of ions and electrons – that could generate a strong magnetic field and magnetize lunar rocks near the impact zone.

NASA is planning to study this mystery in their Artemis mission.

Note: The research team, led by John A. Tardun have William R. Kenan Jr., Professor, of Geophysics and the dean of research for Arts, Sciences & Engineering at the University of Rochester, researchers from Rochester's Department of earth and Environmental Sciences, the Planetary Science Institute (PSI), Michigan Technological University (MTU), the Geological Survey of Japan, and universities across the US, UK, Canada, and France)

Reference

1. *Tarduno, John. Moon lacked a magnetic field for nearly all its history – new research resolves mystery sparked by rocks brought back on Apollo. The Conversation. August 4, 2021.* https://theconversation.com/moon-lacked-a-magnetic-field-for-nearly-all-its-history-new-research-resolves-mystery-sparked-by-rocks-brought-back-on-apollo-163243

Q109. Do magnets work on the moon?

A109. Yes. Either permanent magnets or electromagnets will work on the moon. Magnets are not dependent upon the magnetic field or gravitational field. They are magnetic because of the alignment of the polarized atoms in the metallic crystals of the metal contained in the magnet. Each atom generates a tiny electromagnetic force, and if enough of the atoms are aligned, the magnetic force is sufficient to attract other metallic (mostly iron, nickel, or cobalt-based) objects. However, the magnet in a compass only responds to the magnetic field of the planet or moon that the compass is nearby. In the absence of a planetary magnetic field a compass will not point north. It may instead point towards a local electromagnetic field, like that generated by certain machines or electrical devices.

Q110. Is there a risk of astronauts getting electric shocks on the moon's surface?

A110. The moon's surface gets charged from constant exposures of charged particles from solar winds. earth's magnetic field protects us from these highly charged particles, but the moon lacks a magnetic field.

In a recent study conducted by Joseph Wang and Ziyu Huang of the University of Southern California, the abrasive moon dust, accumulated on astronaut's space suit, can act as pathway for the flow of electrons to the astronauts, which could lead to an electric shock.

Charging was not considered a serious risk during the Apollo mission because astronauts always stayed under sunlit areas. Sunlight itself doesn't carry a charge, but photons hitting the moon's surface can kick up electrons (photoelectric emission) leaving behind a positively charged surface, which balances out the negative charges brought along by the solar wind's electrons. With the two charges more balanced, shocks are less likely in sunlit areas. However, future lunar missions plan to land humans on the moon's south pole where there is little sunlight and high voltage charging can occur. This raises concerns on possible charging/arcing risks for astronauts and equipment/instruments.

References

1. *Wendel, JoAnna. Why the Next Lunar Astronauts May Have to Worry About Electric Shocks. GIZMODO. December 18, 2019.*
https://gizmodo.com/why-the-next-lunar-astronauts-may-have-to-worry-about-e-1840489049
2. *Wang, Joseph. Huang, Ziyu. Dusty Spacesuit Charging/ Arcing in Plasma: Implications for Astronaut Safety at the Lunar Terminator and Far-Side. AGU Fall meeting P33C-09. December 11, 2019*
https://agu.confex.com/agu/fm19/meetingapp.cgi/Paper/586063

❑

CHAPTER 8

Chemicals on the Moon

Q111. What chemical elements are found on the moon?

A111. Randy Korotev of Washington University in St. Louis has listed major and minor elements on the moon based on data from dozens of literature sources and his own lab in samples from the 6 Apollo mission and 3 Russian Luna missions that brought samples back from the moon.

Major Elements in lunar rocks and soils: 99 per cent of the mass consists of the following 7 chemical elements

Elements	**per cent**
Oxygen	41-45 (some reports up to 60)
Silicon	16-17
Aluminium	6-10
Calcium	4-6
Magnesium	3-6
Iron	2-5
Titanium	1-2

Oxygen is the most abundant element on the moon even though there is no air on the moon. The reason is that oxygen bonds to so many things.

Minor Elements: Nearly all the remaining 1 per cent consists to these 4 chemical elements: Manganese, Sodium, Potassium, and Phosphorus

Korotev included the moon soil (regolith) samples and those lunar meteorites that are breccias. The lunar meteorites come from all over the moon, whereas the Apollo and Luna mission samples come from a small area of the nearside of the moon.

Terrestrial geochemists like to express the measured concentration of elements their oxides such as reporting Silicon (Si) as Silicone oxide (SiO2).

Lunar Surface Chemical Composition is given below.

Lunar Surface Chemical Composition

Compound	Formula	Composition	
		Maria	Highlands
Silica	SiO2	45.4%	45.5%
Alumina	Al2O3	14.9%	24.0%
Lime	CaO	11.8%	15.9%
Iron (II) oxide	FeO	14.1%	5.9%
Magnesia	MgO	9.2%	7.5%
Titanium dioxide	TiO2	3.9%	0.6%
Sodium oxide	Na2O	0.6%	0.6%
		99.9%	**100.0%**

Source: Wikipedia

Reference

Korotev, Randy L. Department of earth and Planetary Sciences, Washington University in St. Louis. The Chemical Composition of Lunar Soil.

https://sites.wustl.edu/meteoritesite/items/the-chemical-composition-of-lunar-soil/

Q112. What elements are lacking in the moon rocks?

A112. The lunar crust is greatly depleted in many elements. These fall into several categories: volatile, siderophile, and chalcophile.

Volatile elements have low boiling points and vaporize easily. They include hydrogen and the noble gases (helium, neon, and so on.)

Siderophile (iron-loving) elements tend to alloy in iron metal. Representatives are cobalt, nickel, and the precious metals.

chalcophile (sulphur-loving) elements are those that prefer to combine with sulphur; besides sulfur itself, they include metals such as zinc, lead, and copper.

Q113. What are the primary minerals on the moon?

A113. The main minerals on the moon are:

Plagioclase (aluminium silicates of which anorthite -- calcium aluminium silicate -- is the most common plagioclase mineral)

Olivine (predominantly magnesium and iron silicates)

Ilmenite (iron titanium oxide). It is one of the most valuable minerals on the moon because it is a titanium ore. Some lunar rocks have 10 times as much of the stuff as rocks here on earth do.

Pyroxenes, a chemically complex group of silicate minerals with many variations.

There are other valuable minerals and glasses on the moon. **Rare Earth Metals** (REM) are a good example. REM are important for modern electronics, batteries, and electric motors.

Anorthite is an attractive lunar aluminium resource. The lunar crust contains a much higher proportion of anorthite than does the earth's crust and the lunar highlands are particularly rich in anorthite. Anorthite would also create calcium as a byproduct, since anorthite is a calcium-aluminium silicate. Calcium is the fourth most abundant element in the lunar highlands. Calcium oxides and calcium silicates are not only useful for ceramics, but pure calcium metal is an excellent electrical conductor.

Note: A mineral is an element or chemical compound that is normally crystalline and that has been formed because of geological processes. Minerals are naturally occurring inorganic substances with a definite and predictable chemical composition and physical properties. Minerals made of only one element are called native minerals (i.e., silver, gold, mercury, copper). Minerals composed of compounds are simply referred to as minerals (i.e., quartz, which is made up of silicon and oxygen).

Glass is a non-crystalline, often transparent amorphous solid.

Q114. Is there a relationship between lunar mineral distributions and lunar geology?

A114. In a recent research paper (2021), sun, Lucey, and Taylor (University of Hawaii) have concluded that there is a relationship between global lunar mineral distributions and the geologic evolution history of the moon. Using the results of Apollo 15, 16, and 17 samples as reference, the researchers have mapped global distribution of moon's minerals with data from Kaguya, a Japanese lunar satellite. They have concluded that low-calcium pyroxene is much more abundant than olivine in the lunar highlands and within the South Pole-Aitken basin. Their models suggest that the South Pole Atkin basin excavated deeper and more exposed lunar mantle material that is dominated by pyroxene. However, in contrast, basins and the highlands of large basins, formed after South Pole Atkin, show higher olivine to pyroxene ratio values, may reflect a contribution from olivine-rich upper mantle components.

Reference

Sun, L., Lucey, P. G., and Taylor, G. J. (2021). Correlating Apollo soil mineralogical data with Kaguya spectral data for a global mineralogical classification. Journal of Geophysical Research: Planets, 126(5), e2020JE006445. DOI: https://doi.org/10.1029/2020JE006445

Q115. Why is the moon depleted in volatile elements?

A115. According to the Giant Impact Theory, the moon formed about 4.5 billion years ago when a Mars size object, known as Theia, collided with the early proto earth. It is therefore expected that volatile elements, found on earth such as sodium, potassium, and zinc, to be present in the moon rocks. However, Lunar rocks lack such volatile elements. So, why do we not find the volatiles in lunar rocks samples has always been a question in people's minds?

Dr. Robin Canup and colleagues (Southwest Research Institute, USA) studied this by combining new and existing models of how the moon formed. The team has hypothesized that the core of the moon initially coalesced from material in the outer part of the disk of debris left over from the collision. This part was sufficiently cool for volatile elements to condense. But the outer layers of the moon came from melt material from the inner part of the debris disk which was still too hot for volatile elements to condense from vapor to solid. The inner disk eventually cooled to temperatures at which volatile elements can condense, but this occurred after the moon had moved away from the debris disk and its formation has ceased. So, the volatiles end up getting preferentially swept up by the earth rather than by the moon.

References

1. *Gary, Sturt. Mystery of the Moon's missing volatile elements finally solved. ABC News. ABC Science. November 9, 2015*
https://www.abc.net.au/news/science/2015-11-10/moon-missing-volatile-material-mystery-solved-lunar/6918918

2. *Canup, R., Visscher, C., Salmon, J. et al. Lunar volatile depletion due to incomplete accretion within an impact-generated disk. Nature Geosci 8, 918–921 (2015).*
https://doi.org/10.1038/ngeo2574
https://www.nature.com/articles/ngeo2574

Q116. Which chemicals and mineralogical resources of the moon can be used?

A116. The moon bears substantial chemical resources (volatiles and minerals) which could be exploited in the future as a potential for in situ resource utilization (ISRU) at the moon or they could be imported to the earth's surface where they would contribute directly to the global economy. This will provide a means of reducing the cost of lunar exploration and beyond.

ISRU refers to the in-situ generation of consumables for autonomous or human activities from raw materials found on the moon. For a successful application of ISRU, it is important that landing site selection, surface operations and suitable ISRU technologies are identified using knowledge of the availability and distribution of lunar resources and detailed understanding of the workings of the various processes available.

Here are some of the chemical and mineralogical resources on the moon which can be used: Oxygen, Water, Hydrogen, Iron, Titanium, Aluminum, Silicon, Calcium, Magnesium, KREEP - means Potassium (K), Rare earth Elements (REE) and Phosphorus (P), Helium-3, Precious Metals.

❑

CHAPTER 9

Helium-3 on the Moon

Q117. What is Helium-3 (He3) and why is it so important?

A117. Normal Helium, which is used to blow up balloons, has 2 protons and 2 neutrons in its nucleus, giving it an atomic weight of 4. We can call it Helium-4. If you kick out one of the neutrons from Helium-4, you get Helium-3, with 2 protons and 1 neutron in nucleus. Therefore, Helium-3 is a lightweight isotope of normal Helium.

Helium-3 could provide safer nuclear energy. It has a potential to be used as a fuel in future nuclear fusion power plants. Experts believe that 1 gram of Helium-3 can replace over 100 tons of coal.

There is very little helium-3 available on the earth. However, there is a significant supply on the moon. Several nations have subsequently signaled their intention to go to the moon to mine helium-3 as a fuel supply. Such plans may come to fruition within the next two to three decades and may trigger a new Space Race.

> **Note:** Isotopes are atoms of the same element with the same number of protons but different numbers of neutrons.

Q118. How did the moon acquire Helium-3?

A118. The sun produces vast quantities of Helium by nuclear fusion of hydrogen atoms. It takes four hydrogen atoms to fuse into each helium atom. This fusion reaction provides energy to the sun.

In the core of the sun, about 600 million tons of hydrogen produces 596 million tons of Helium per second. The remaining 4 million tons per second gets converted to energy (remember the famous Einstein equation E=MC^2). A small amount of Helium-3 (He3) is also produced. Both types of helium travel towards earth as part of the solar wind. He3 never makes it to earth because the earth's magnetic field pushes it away. The moon does not have a big magnetic field. He3 gets accumulated in the debris layer of moon's dust and rock.

Note: The sun has used up about half of its hydrogen fuel in the last 4.6 billion years since its birth. It still has enough hydrogen to last about another 5 billion years.

Q119. How much Helium-3 (He3) is there on the moon?

A119. It is estimated that a total of 1.1 million metric tons of He3 have been deposited in moon-soil. At the current rate of energy consumption this will take care of 10,000 years of world-wide demand. Thus, Helium-3 could power earth's needs for centuries. Helium-3's value is estimated to be at about 3-5 billion dollars a ton, meaning the moon has trillions of dollars' worth of He3. That is why a serious interest is being taken in lunar Helium-3.

Reference

1. *Barnatt, Christopher. Helium-3 Power Generation. Explaining the Future.com*
https://www.explainingthefuture.com/helium3.html

2. Grossman, David. Space Force: the truth behind the helium-3 plot twist. Inverse.com, June 3, 2020. *https://www.inverse.com/innovation/space-force-helium-3*

Q120. Why is Helium-3 a better nuclear fuel?

A120. First let us understand how we get energy from a nuclear reactor.

An atom has a nucleus with positively charged protons and neutrons. Both protons and neutrons are known as nucleons. Negatively charged electrons circle around the nucleus in different energy level orbits. The nucleons are very tightly bound together inside a nucleus.

While, chemical reactions take place by trading and sharing electrons among atoms, nuclear reactions involve the rearrangement of nucleons in the nucleus. Such rearrangements produce energy in a big way based on Einstein's famous equation E=MC2. This is the energy that the nuclear reactors produce which turn water into steam which drives a turbine to produce electricity. Some nuclear reactors split up an atom's nucleus (fission reaction) to produce energy, while others fuse or combine the nucleons (fusion reaction) to produce energy. Currently, all nuclear reactors are fission type as humans are yet to gain complete technological control over the fusion process for a peaceful use.

A fission reactor produces radioactive waste as a by-product which must be safety stored, effectively indefinitely. When things go out of control, hazardous radioactive substances may get released into the environment. On the other hand, nuclear fusion is safe. Nuclear fusion effectively makes use of the same energy source that fuels the sun and other stars and does not produce radioactivity and nuclear waste.

There are some experimental/test/prototype nuclear fusion reactors in the world. In these hydrogen isotopes, tritium (containing one proton and two neutrons in its nucleus, also known as hydrogen 3) and deuterium (containing one proton and one neutron in its nucleus, also known as heavy hydrogen) are used as fuel. Atomic energy gets released when their nuclei fuse to create helium and a neutron. The so-termed "fast" neutrons released lead to significant energy loss and are extremely difficult to contain an automatic runaway process. One potential solution is to use helium-3 and deuterium as fuels in fusion reactors. The involved nuclear reaction here creates normal helium and a proton, which wastes less energy and is easier to contain. Nuclear fusion reactors using helium-3 could therefore provide a highly efficient form of nuclear power with virtually no waste and no radiation.

D + He3 -------→ He4 + H + ~18 MeV

Deuterium + Helium-3 -------→ Normal Helium (or Helium 4) + Proton + ~18MeV

(1P+1N) + (2P+1N) -------→ (2P+2N) + 1P + ~18MeV

P = Proton

N = Neutron

References

1. *Bennett, Gregory. Will Helium-3 become a fuel source for the twenty-first century? NuEnergy. February 7, 2013* https://www.nuenergy.org/will-helium-3-become-a-fuel-source-for-the-21rst-century/
2. *Bonde, Jeff & Tortorello, Anthony. Helium-3, The energy source of tomorrow.* https://mdcampbell.com/Helium-3version2.pdf

Q121. What are the challenges in using the moon's helium-3 for generating power?

A121. There are numerous obstacles to overcome before Helium-3 can be used to generate power. Some of the challenges are:

1. Obtaining helium-3 from lunar regolith will not be an easy task. The best estimates of Helium-3 content place it at 50 parts per billion in lunar soil, calling for the refining of millions of tons of lunar soil before gathering enough Helium-3 to be useful in fusion reactions on earth. One million tons of moon-soil will have to be mined and processed for every 70 tons of helium-3 yield.
2. After mining lunar rock, Helium-3 is separated by heating the mass to over 600 degrees Celsius (over 1100 Fahrenheit), consuming a large amount of energy in the process.
3. Transporting large quantities of Helium-3 back to earth will be another problem. A spacecraft would likely be able to only carry a few tons of Helium-3 as payload. Some suggest having nuclear plants on the moon's surface. In that case transporting energy to the earth needs to be worked out.
4. The temperatures required to achieve helium-3 fusion reactions are much higher than in traditional fusion reactions.

Scientists and engineers all over the world are busy in developing technologies and engineering to make things workable.

❑

CHAPTER 10

Water on the Moon

Q122. Is water available on the moon?

A122. Yes, there is water on the moon. At the time of the Apollo and Luna programme, scientists were not quite well equipped to detect the presence of water in the lunar rock samples. Now, with the improved testing facilities, they have established that those rock samples contain measurable amounts of water, chemically locked with the molecules of the rocks and dust. Water has been detected in lunar rocks, soil as well as in a very thin atmosphere. Water creation, migration, deposition, and retention are occurring on the moon's surface. Chandrayaan-1 mission confirmation of water on the moon in 2008 has created a lot of interest and a great amount of "water exploration on the moon" work is going on.

Q123. How and when was the presence of water on the moon confirmed?

A123. The first indication of the presence of water on the moon came in 1978 when Soviet Luna 24 moon dust (Regolith) data showed 0.1 per cent water by mass. However, by 1978 it was widely held that the Apollo samples contained virtually no water, The Soviet findings were completely ignored. Later scientists learned that the no-water result of Apollo samples was seriously mistaken. While that debate was continuing, the definitive discovery of water on the moon came in 2008 when the Indian Space Research Organization's (ISRO's) Chandrayaan-1 was sent to lunar orbit. Chandrayaan-1 discovered the presence of water in the moon's atmosphere as well as on its surface.

1. Discovery of water in the moon's atmosphere

Chandrayaan-1 included a lunar orbiter and an impactor - moon Impact Probe (MIP). MIP carried a device named Chandra Altitudinal Composition Explorer (CHACE), developed to detect molecules of elements present in the moon's atmosphere. CHACE was switched on just fifteen minutes before the MIP separation. As long as MIP was under free-fall towards the moon's surface, CHACE was mapping whatever trace constituent elements that were present in the moon's extremely thin atmosphere. The scientists at ISRO were getting data live. Within minutes they knew they had found water molecules floating around in the moon's atmosphere.

2. Water on the moon's surface

The MIP was designed to collide with the moon's surface and throw up some underground soil as ejecta which could then be remotely analysed by the instruments onboard the orbiting Chandrayaan-1 for any signs of presence of water ice, and other materials. The MIP impacted Shackleton Crater. On September 25, 2009, ISRO announced that the MIP had indeed helped discover water on moon's surface.

Chandrayaan-1 also carried onboard NASA's moon Minerology Mapper (M3) and Mini-SAR (Mini Synthetic Aperture Radar) instruments. Chandrayaan-1 was placed at a height of 100 km from the moon's surface. M3 and Mini-SAR continued studying the moon from that distance. M3, not only confirmed the existence of water, but it also found that water signal got stronger toward the polar regions. Mini-SAR also detected ice deposits near moon's poles. The instrument has imaged many of the permanently shadowed regions which are extremely cold and water ice could be stable there indefinitely.

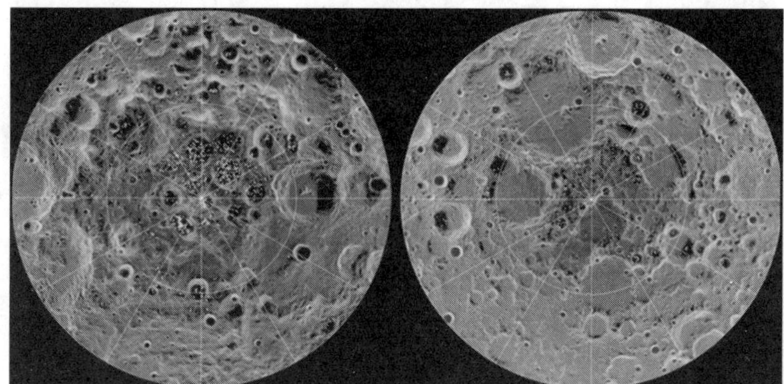

The image shows the distribution of surface ice at the moon's south pole (left) and north pole (right), detected by NASA's moon Mineralogy Mapper instrument on board in ISRO's Chandrayaan-1 spacecraft. Blue represents the ice locations, plotted over an image of the lunar surface, where the grey corresponds to surface temperature (darker representing colder areas and lighter shades indicating warmer zones). The ice is concentrated at the darkest and coldest locations, in the shadows of craters. This is the first-time scientists have directly observed definitive evidence of water ice on the moon's surface.

Credits: NASA

https://www.nasa.gov/feature/ames/ice-confirmed-at-the-moon-s-poles

Q124. Are there more confirmations of water on the moon after the Chandrayaan-1 mission?

A124. Yes, there are more confirmations of water on the moon after the Chandrayaan-1 mission.

1. **Spacecraft Cassini** passed by the moon in 1999 on its way to Saturn. It has confirmed stronger signals of water. The data shows a global distribution of the water signal, though it appears stronger near the poles.

2. **LRO and LCROSS**

In 2009, NASA launched the Lunar Reconnaissance Orbiter (LRO) and an impact probe LCROSS (Lunar Crater Observation and Sensing Satellite). The upper stage from the rocket used to launch LRO and LCROSS was deliberately smashed into a dark crater, Cabeus, near moon's southern polar region. LCROSS flew through the debris plume and detected a significant amount of hydroxyl group in the material thrown up before it crashed into the moon.

3. SOFIA

In 2020, SOFIA's (Stratospheric Observatory For Infrared Astronomy) telescope confirmed the presence of water on the sunlit, non-polar surface of the moon. This discovery indicates that water may be distributed across the lunar surface, and not limited to cold, shadowed places. SOFIA is an airborne observatory, a joint project of NASA and the German Aerospace Center's (DLR).

Since then, many other moon missions have seen hints of water in different forms and in different regions.

4. Chang'e 5

More recently, China's Chang'e-5 lander has directly detected water in the regolith under its feet. While it found low levels of hydroxyl (OH) in the regolith all around it, one rock, in particular, had almost twice the concentration. It's believed this rock came from a different, more water-rich region and was blasted by a meteor impact millions of years ago.

Reference

Atkinson, Nancy. China's Lander Has Detected Water on the Moon. Universe Today. January 12, 2022.

https://www.universetoday.com/154006/chinas-lander-has-detected-water-on-the-moon/

Q125. How can water exist on the moon when it has no atmosphere and daytime temperatures reach 120 degrees Celsius (250 degrees Fahrenheit).

A125. Unlike Our earth, the moon has less of axial tilt (earth's tilt is 23 degrees while moon's is 1.5 degrees). At the moon's poles the sun hovers close to the horizon. Large craters with terraced rims block sunlight from ever reaching inside. These areas are called Permanently Shadowed Regions (PSRs). Temperatures inside the moon's PSRs can drop to -250 degrees Celsius (-418 degrees Fahrenheit). That is why water ice stays there. The moon's poles have hundreds of PSRs where substantial water ice has been detected.

Q126. Where did water on the moon come from?

A126. There are potentially two types of water on the moon:

1. Exogenic - water from outside sources, such as comets striking the moon's surface, and
2. Endogenic - water that originates on the moon.

There are several hypotheses about how water is formed on the moon, some of which are given below:

1. Water brought from asteroids and comets:

Water found on the moon's surface are the result of striking water-bearing asteroids and comets. Researchers think that there was a time period when earth and the moon were being bombarded by comets and asteroids, which contain a lot of water ice, and this might have delivered to the earth's oceans and lakes as well as the moon's icy deposits. Using computer models, Russian scientists have concluded that only a small amount of water that arrives with comets stays on moon's surface compared to what is brought by asteroids.

2. Water formed on the Moon's surface after exposure to the solar wind

The sun's outer layer (known as corona) reaches extremely high temperatures up to 1.1 million Celsius. At such a high temperature, the sun's gravity cannot hold on to the rapidly moving particles and they stream away as solar winds. The material of the solar wind would do severe damage to any life that might exist. We are fortunate because earth's magnetic field serves as a shield, redirecting the material around earth. The moon, on the other hand, has only traces of a non-uniform magnetic field, therefore its surface is bombarded with solar wind. When the charged high-speed hydrogen particles in solar wind hit the moon's surface, a chemical reaction gets triggered with oxygen atoms in lunar soil and water is formed. Moon's soil samples brought by Apollo and Luna missions have now confirmed that they are enriched in atomic nuclei deposited from the solar wind.

Recently, Japan's scientists have demonstrated this in the lab experiments. These results will help to estimate the amount of water on the moon.

3. Water from earth wind (water bridge idea)

A new study conducted by H.Z. Wang and others proposes that solar wind is not the only source of water-forming ions on the moon but some of the water on the moon's surface came from earth's wind as well.

Earth's magnetic field was thought to block the solar wind from reaching the moon during full moon period (3 to 5 days in each lunar cycle). During this period, water could not be regenerated faster than it was lost due to evaporation. The researchers found this was not the case. Based on data from India's Chandrayaan-1 satellite's moon Mineralogy Mapper and by comparing a time series of water surface maps before, during and after the magnetosphere transit, the researchers argue that lunar water could be replenished by flows of magnetospheric ions, also known as "earth wind." The presence of these earth-derived ions near the moon was confirmed by Japan's Kaguya satellite.

The authors have suggested that future studies may provide further evidence that there's a water bridge between the earth and the moon. China's Chang'e 5 mission returned lunar samples to earth back in December. Those samples could hold evidence for the water bridge idea.

References

1. *Mehta, Jatan. Your Guide to Water on the Moon. The Planetary Society. November 23, 2020* https://www.planetary.org/articles/water-on-the-moon-guide

2. *Gough, Evan. The Earth's Magnetosphere Might be Creating Water on the Moon. Universe Today. February 3, 2021*
https://www.universetoday.com/149945/the-earths-magnetosphere-might-be-creating-water-on-the-moon/
3. *Wang, H. Z. et al. earth Wind as a Possible Exogenous Source of Lunar Surface Hydration. The American Astronomical Society. April 907 L32. January 28, 2021.*
https://iopscience.iop.org/article/10.3847/2041-8213/abd559
4. *Masongsong, Emmanuel (UCLA earth, Planetary, and Space Sciences). First Evidence That earth's Magnetosphere Can Create Water on the Moon's Surface. Phys.org. January 28, 2021*
https://phys.org/news/2021-01-evidence-lunar-surface-earth-magnetosphere.html

Q127. How much water is estimated on the moon?

A127. The short answer to the question is: We are not sure. There are several estimates given in the literature and work is in progress. Some of the estimates are given below:

1. Map of water in the uppermost portion of the moon's soil

Based on data taken from NASA's moon Mineralogy Mapper (M3), which flew aboard India's Chandrayaan-1 spacecraft, scientists from Brown University have quantified, as a road map, how much water is present on a global scale. They have created the first quantitative map of water and its chemical building blocks trapped in the uppermost portion of the moon's soil. This has been discussed by Kevin Stacey in a featured events news from Brown University.

The scientists have concluded:

(a) The distribution is largely uniform rather than splotchy, with concentrations gradually decreasing towards the equator.

(b) The amount of water does not show a significant difference among distinct compositional terrains.

(c) The water concentration reaches a maximum average of around 500 to 750 parts per million in the higher latitudes.

(d) Higher-than-average concentrations of water was found in lunar volcanic deposits near the moon's equator, where background water in the soil is scarce. Rather than coming from solar wind, the water in those localized deposits likely comes from deep within the moon's mantle and erupted to the surface in lunar magma.

(e) The concentration of water changes over the course of the lunar day going from wetter in the early morning and evening to much drier around lunar noon. The fluctuation can be as much as 200 parts per million. Although the exact mechanism for this fluctuation is not known, it tells that the process of water formation in the lunar soil is active and happening today.

(f) It is also not clear how deep into the soil the water mapped in the study goes. The distribution of water with depth could make a big difference in terms of how much water is actually there.

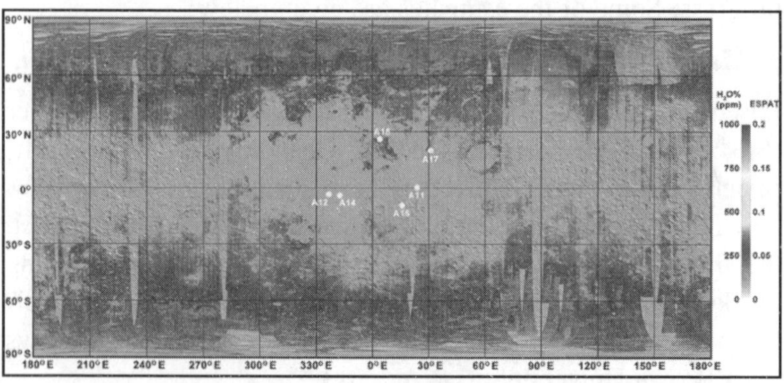

The map shows a general trend of increasing water content (purple, green and yellow) toward the poles. The dots mark the Apollo landing sites.
Credits: Shuai Li / Ralph Milliken / Brown University
https://www.brown.edu/news/2017-09-13/moonwater
Reproduced with permission

2. Data from Chandrayaan-1 and LRO (Lunar Reconnaissance Orbiter)

Based on the instruments aboard Chandrayaan-1 and Lunar Reconnaissance Orbiter (LRO), the lunar poles have over 600 billion kilograms of water ice. This is a low-end estimate because the amount of water detected is limited by the strength of spacecraft radar. New missions with radars that penetrate deeper will likely find more water ice.

In 2020, researchers, using LRO data, also identified Permanent Shadowed Regions (PSRs), inside smaller craters. They may be responsible for housing 10 to 20 per cent of the water stored in all the moon's permanent shadows, for a total area of about 40,000 square kilometres (15400 square miles), mostly in regions closer to the poles.

3. Data from LCROSS (Lunar Crater Observation and Sensing Satellite)

Data from LCROSS indicate an abundance of about 5.6 ± 2.9 per cent.

4. Data from SOFIA (Stratospheric Observatory For Infrared Astronomy)

According to the observations from SOFIA, the amount of water is 12 ounces dispersed over a cubic metre (340 grams per cubic metre) of lunar soil. According to another estimate, one ton of soil would hold about 32 ounces (900 grams) of water, which will be about 1500 grams per cubic metre.

In conclusion, there is water on the moon, not much but the moon is not dry. More work is in in progress about "Water on the moon".

References

1. *Stacey, Kevin. Brown University. Researchers create first global map of water in moon's soil. News from Brown – Science and Technology. September 13, 2017* https://www.brown.edu/news/2017-09-13/moonwater

2. *Patel, Neel V. Water on the moon should be more accessible than we thought. MIT Technology Review. October 26, 2020*
https://www.technologyreview.com/2020/10/26/1011228/water-moon-more-accessible-nasa-sofia-lro-glass-micro-cold-trap/

Q128. How will future missions study water on the moon?

A128. Several missions, as outlined below, are lined up to study water on the moon. Scientists are learning more and more about Permanently Shadowed Regions (PSRs) where more ice is available. Extensive maps of LRSs have been prepared.

1. ISRO's Chandrayaan-2 orbiter, launched in 2019, is using its advanced radar to map the water ice at greater depths and quantify it. These efforts will be aided by NASA's Lunar Flashlight CubeSat, planned to be launched in 2022 as a part of Artemis-1 mission payloads.

2. The South Korea's first moon orbiter, Pathfinder Lunar Orbiter (KPLO), is expected to launch in mid-2022. This will carry a US-built instrument ShadowCam. ShadowCam will peer inside dark craters near the moon's poles, where previous missions detected evidence of water ice deposits.

3. NASA's lunar rover, the Volatiles Investigating Polar Exploration Rover (VIPER) will explore the moon in search of water ice and other potential resources. This mobile robot will land at the South Pole of the moon in late 2023 on a 100-day mission. The critical information obtained will teach us about the origin and distribution of water on the moon and help determine how we can harvest the moon's resources for future human space exploration.

4. The European Space Agency (ESA) is working on a special rover that would be powered by a laser, to operate in the darkness of these PSRs to search for water. The laser power source would itself be solar powered.

The lunar poles are also central to human and robotic exploration plans of commercial companies and many other nations, including China, India, Japan, Europe, and Russia.

Q129. What are the challenges in exploring the moon's water ice directly?

A129. The first hurdle is developing an ability to land with precision. Whereas regions near the moon's equator have large portion of flat terrain, the polar regions are often rocky and sloped. The lunar poles also require spacecraft that can survive very cold temperatures. While PSRs are extremely cold, their sunlit crater rims don't get very much direct sunlight, keeping them cooled to an average of about -50 degrees Celsius (-58 degrees Fahrenheit). Another challenge is power and communication. A rover venturing into a PSR will lose its line-of-sight with the earth, disabling communications, while the lack of sunlight means it will have to pack powerful batteries or rely on nuclear power.

One of the ideas, being looked at by NASA is to use beamed power, transmitting energy from an illuminated tower to a rover in the darkness.

References

1. *Atkins, Nancy. Yes, There's Water on the Moon. Universe Today – Space and Astronomy News. September 23, 2009* https://www.universetoday.com/41212/yes-theres-water-on-the-moon/
2. *Cane, Frazer. Dust Off Your Lunar Colony Plans. There's Definitely Ice at the Moon's Poles. Universe Today. August 23, 2018.* HTTPS://WWW.UNIVERSETODAY.COM/139819/DUST-OFF-YOUR-LUNAR-COLONY-PLANS-THERES-DEFINITELY-ICE-AT-THE-MOONS-POLES/

3. *Chou, Felicia. Hawkes, Alison. NASA's SOFIA Discovers Water on Sunlit Surface of Moon. Press Release-Earth's Moon. October 26, 2020*
https://www.nasa.gov/press-release/nasa-s-sofia-discovers-water-on-sunlit-surface-of-moon
4. *Cohen, Barbara A. Hayne, Paul O. Greenhagen, Benjamin. Paige, David A. Seybold, Calina. Baker, John. Lunar Flashlight: Illuminating the Lunar South Pole. IEEE Aerospace and Electronic Systems Magazine Volume: 35, Issue: 3. March 1, 2020.*
https://ieeexplore.ieee.org/document/9076196/figures#figures
5. *Crotts, Arlin. Water on the Moon. The Space Review. October 12, 2009 (This paper has reference of Luna 24)*
https://www.thespacereview.com/article/1485/1
6. *Gouch, Evan. Universe Today. This Laser Powered Rover Could Stay in the Shadows on the Moon and Continue to Explore. May 19, 2020*
https://www.universetoday.com/146118/this-laser-powered-rover-could-stay-in-the-shadows-on-the-moon-and-continue-to-explore/
7. *Hecht, Jeff. Water on the Moon isn't Just at the Poles. Sky & Telescope. October 26, 2020*
https://skyandtelescope.org/astronomy-news/water-on-moon-not-just-poles/
8. *Mehta, Jatan. Interviewing Chandrayaan 1's Mission Director on India's Role in the New Moon Race. Jatan-Space. August 7, 2020*
https://blog.jatan.space/p/interviewing-isro-chandrayaan-1-mission-director
9. *Mehta, Jatan. What SOFIA's discovery of water on the Moon is and isn't. Jatan-Space. October 27, 2020.*
https://blog.jatan.space/p/on-sofias-discovery-of-water-on-the-moon
10. *Soderman/NLSI Staff. SERVI (Solar System Exploration Research Virtual Institute). It's Official: Water Found on the Moon. 2009.*

https://sservi.nasa.gov/articles/its-official-water-found-on-the-moon/
Main Source: http://www.space.com/scienceastronomy/090923-moon-water-discovery.html

11. *Stacey, Kevin (Media Contact). Study suggests ice on lunar south pole may have more than one source. Brown University (Science and Technology). October 10, 2019.*
https://www.brown.edu/news/2019-10-10/moonice
12. *Williams, Matt. Lunar Landings Will Make it Harder to Study the Moon's Ice Deposits. Universe Today. August 21, 2020.*
https://www.universetoday.com/147512/lunar-landings-will-make-it-harder-to-study-the-moons-ice-deposits/
13. *Williams, Matt. Mapping Out the Water on the Moon. Universe Today. April 2, 2020.*
https://www.universetoday.com/145527/mapping-out-the-water-on-the-moon/
14. *Mehta, Jatan. Ultimate guide to water on the Moon. Jatan's Space. November 26, 2020.*
https://blog.jatan.space/p/ultimate-guide-to-water-on-the-moon?utm_source=url

❑

PART II: MOON IN MOTION

"Go slowly, my lovely moon, go slowly."

— Khaled Hosseini

CHAPTER 11

Moon in Motion (Orbital and Rotational)

Moon's Orbital Motion

Q130. Why does the moon orbit around the earth?

A130. Orbits are the result of a balance between the forward motion of a body in space (inertia) and the pull of gravity on it from another body in space. An object with a lot of mass goes forward and wants to keep going forward (Newton's first law of motion). However, the gravity of another body in space pulls it in. There is a continuous tug-of-war between the one object wanting to go forward and away and the other wanting to pull it in.

These forces of inertia and gravity must be perfectly balanced for an orbit to happen. If the forward movement (inertia) of one object is too strong, the object will speed past the other one and not enter an orbit. If inertia or momentum is much weaker than the pull of gravity, the object will be pulled into the other one completely and crash.

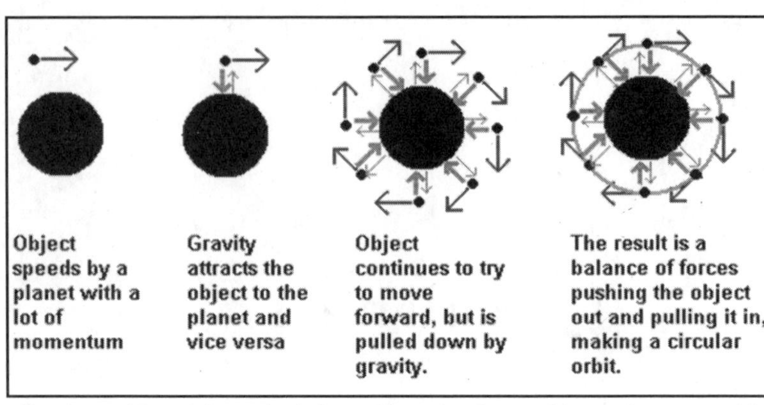

Gravity and Inertia must be balanced in order for an object to be captured in an orbit around another object.
https://www.qrg.northwestern.edu/projects/vss/docs/space-environment/1-what-causes-an-orbit.html
Reproduced with the permission from www.qrg.northwestern.edu

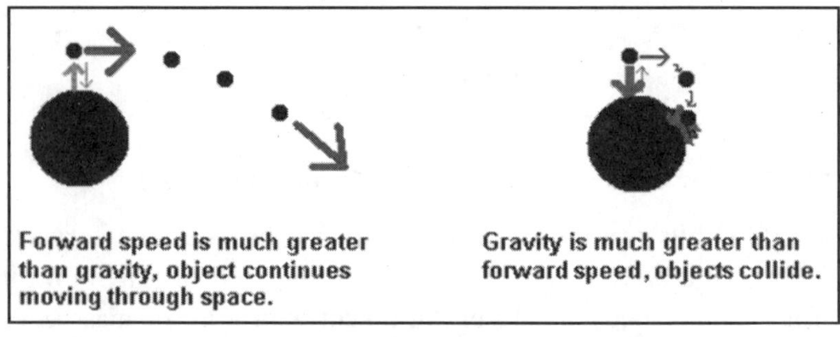

Gravity and Inertia

Reproduced with the permission from www.qrg.northwestern.edu

Reference

QRG-Northwestern-EDU. What causes an orbit to happen? https://www.qrg.northwestern.edu/projects/vss/docs/space-environment/1-what-causes-an-orbit.html

Q131. Why do objects in space follow elliptical orbits?

A131. The orbit of an object around its 'parent' is a balance between the force of gravity and the inertia (object's desire to move in a straight line. To fully understand orbit, we need to consider the energy that both objects bring into the system, and the effects that will have on the shape of the orbit.

The shape of an orbit is from the fact that the force of gravity between two objects depends on the square of the distance between them. If we double the distance between two objects, the attractive force between them drops to a quarter of its original value, if we triple the distance it drops to a ninth, and so on. Isaac Newton demonstrated mathematically (using calculus) that this inverse square law implied that the path followed by an object in a gravity field would be a parabola, a hyperbola, or an ellipse. The first two are open ended. If something entered in a parabolic or hyperbolic path, we would see it just once before it disappeared into the distance. The Ellipses are closed so the bodies (planets and moons) we see in elliptical orbits stick around. A circle is a special case of an ellipse, and it is theoretically possible for an orbit to be circular. In the real world, such an orbit is unlikely.

> **Note:** Prior to Newton, Kepler showed by measurement that the observable planets and moons had elliptical orbits. But he did not know how to explain it mathematically. Newton invented calculus to prove it.

Q132. The sun is larger than earth and has more gravitational force, then why the moon does not orbit around sun instead of earth?

A132. You might think this is because the earth is much closer to the moon than the sun. But this is not the case. The sun is about 400 times more distant from the moon than the earth, but the sun is about 330,000 times more massive. By using Newton's law of gravity, one can find that even with its greater distance, the sun pulls on the moon about twice as hard as the earth does.

So why doesn't the sun steal the moon?

To escape the gravitational pull of a body, you need to be moving fast enough relative to that body to escape its pull. This is known as the escape velocity of the object. The speed of the moon around the earth is about 1 km/s. At the moon's distance from the earth, the escape velocity is about 1.2 km/s. The moon simply isn't moving fast enough to escape the earth.

There is another way to look at it, Hill/Roche sphere, the size of the sphere around the earth (or any large object) in which it will have the dominant influence. To be retained by a planet, a moon must have an orbit that lies within the planet's Hill sphere. There is a mathematical equation that we can use to figure it out. The Hill/Roche sphere radius of earth is 1,471,000 km (914,040 miles). So, at 384,400 km, the moon is safely within the earth's Hill/Roche Sphere and thus is securely attached to the earth.

References

1. *Koberlein, Brian. Why Doesn't the sun Steal The Moon? Universe Today. November 10, 2014.* https://www.universetoday.com/116158/why-doesnt-the-sun-steal-the-moon/
2. *Hill sphere. Wikipedia, the free encyclopedia. Last edited on August 9, 2021.* https://en.wikipedia.org/wiki/Hill_sphere

Q133. What are some of the moon's orbital details?

A133. 1. The moon orbits around the earth in an eccentric orbit, and not circular. The orbital eccentricity of an astronomical object is the amount by which its orbit around another body deviates from a perfect circle. A value of 0 is a circular orbit, values between 0 and 1 form an elliptic orbit. The moon's eccentricity is 0.0549, therefore the moon's orbit is elliptical (some people call it an elongated circle). To compare with the earth, the eccentricity of the earth's orbit is currently about 0.0167 which means that the earth's orbit is more circular than the moon's.

2. The moon travels around earth once every 27.322 days, known as a sidereal month. (More on this later)
3. If we looked down on the earth-moon system from the north celestial pole, we would see that the moon travels around the earth in a counter-clockwise orbit.
4. The moon travels in its path around the earth at a speed of 3,680 kilometres per hour (2,287 miles per hour). This is approximately 1 km per second.
5. The moon's distance from earth varies throughout its monthly orbit because the moon's orbit is elliptical. Every month, the moon's elliptical orbit carries it to its most distant point from earth, known as apogee – and then to the moon's closest point to earth, known as perigee – roughly two weeks later.

Perigee (i.e., min. distance from earth): average 363,230 km (225,700 mi)

Apogee (i.e., max. distance from earth): average 405,400 km (251,900 mi)

Mean distance 385,000 km (239,000 mi)

Mean inclination of orbit to ecliptic 5.15° (range 4.99–5.30)

Mean inclination of lunar equator to ecliptic 1.543°

Mean eccentricity 0.0549006 (range 0.026–0.077)

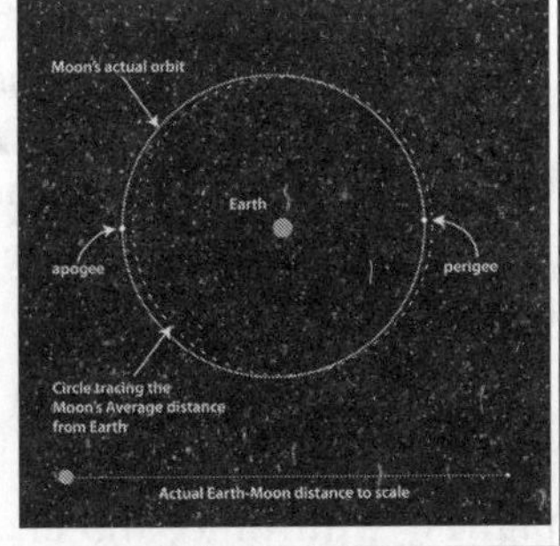

The moon's orbit around earth is not a circle, but it is very nearly circular, as the above diagram shows.
Credit: Diagram by Brian Koberlein.
https://earthsky.org/upl/2015/09/perigee-apogee-brian-koberlein-e1443384711895.png
Reproduced with permission.

Q134. Why does the moon's apparent size change across its orbit, sometimes looking bigger and sometimes smaller (not to confuse with phases of the moon)?

A134. Since nearer objects appear larger and farther appear smaller, the moon's apparent size changes as it moves toward and away from an observer on earth. An event referred to as a 'supermoon' occurs when the full moon is at its closest to earth (perigee). The largest possible apparent diameter of the moon is 12 per cent larger (as perigee versus apogee distances) than the smallest; the apparent area is 25 per cent more.

The illustration, based on Galileo spacecraft images, shows the approximate difference in apparent size between a full moon at perigee (left) and a full moon at apogee, the farthest point in the lunar orbit.
Credits: Wikimedia Commons. Galileo Project, NASA. December 22, 1999
Source: http://apod.nasa.gov/apod/image/9912/twomoon.jpg
https://commons.wikimedia.org/wiki/File:Perigee_moon,_Apogee_moon_%28Galileo%29.jpg
Figure in Public Domain

Q135. Why are the planets and their moons all rotate / orbit in the same plane?

A135. The reason for the planets and their moons to orbit in the same plane is the law of conservation of angular momentum. Even if orbits of each body may begin in different orbital planes, with time, the energy is lost against the sun's direction of rotation. It is like when you spin a ball fast, it starts to flatten from the top.

To understand the answer to the question fully, we must understand how celestial bodies were formed. If we talk about our solar system, this happened more than 4 billion years ago. Scientists believe that our solar system is thought to have arisen from an amorphous cloud of gas and dust in space. The original cloud was spinning. Due to the conservation of angular momentum, as the cloud spins faster and faster, it collapses into a disk, which is a balance between gravitational collapse and centrifugal force created by the rapid spin. The sun, planets, moons, and other celestial bodies are believed to have formed out of this disk, which is why, today, they all still orbit in a single plane around the sun, known as ecliptic. Also, due to the concept of energy minimization, anything that was not in that plane would not last very long. They would be more likely to knock into each other, lose orbital velocity and then fall into the sun.

It is interesting to note that distant stars and planets follow the same pattern because they were also formed from spinning clouds of gas and dust in space.

Interestingly, there are certain exceptions like:

1. In early 2019, astronomers announced a newly discovered double star system where the planetary disk has somehow been kicked up over the expected equatorial plane, and now appears oriented over the plane of the stars' poles.
2. If we go far enough away from our sun (beyond the planets and asteroids, beyond the Halley-like comets and even beyond the Kuiper Belt), bodies are moving like a swarm rather than a nice, orderly set of nearly perfect circles.

Q136. Why does the moon appear higher in the winter nights sky and lower in the summer nights?

A136. Compared to most other planets and their moons, the earth's moon orbit is a bit unusual. Most other moons revolve along the equatorial planes around their parent planets. However, the earth's moon orbit is highly tilted. It makes an angle close to 29 degrees with the plane of equator of the earth. As a result of this orbital tilt, the moon appear higher in the sky in winter nights and again lower in the sky in summer nights.

Scientists have long been baffled by the moon's peculiar incline. They believe that this is due to gravitational tugs the moon received from giant, close-passing rocks, that eventually slammed into the earth. The early solar system was full of such giant rocks hurtling around and later colliding with the earth and the other planets. Based on computer modelling, the researchers found that there is a strong possibility that a few tens of millions of years after moon's formation, these close passes could have given moon's orbit the tilt that we see today.

Q137. Earth's gravity pulls the moon then why doesn't the moon fall into the earth and crash?

A137. In fact, the moon is constantly falling down. Objects in a gravitational field always fall towards the source of gravity. The moon is continually falling towards the earth and missing it. All gravitational orbits are cases of falling and missing, a state of free fall.

There are two ways of looking at this.

(1). The way Newton thought of it:

There is a gravitational force between the moon and the earth. This force pulls the moon towards the earth. This constant tug on the moon as it moves around the earth is called a "centripetal" force. This force is balanced by the "centrifugal" force, that should otherwise see the moon flung off in a straight line because of its orbital velocity (Newtons first law of motion). It is the balance between the centripetal and centrifugal forces that keep the moon in orbit around the earth.

Newton had explained the nature of orbits by a thought experiment. Consider a cannon on the surface of the earth that shoots a cannonball straight forward. As the ball speeds forward, earth's gravity pulls on it and it falls to the earth until it hits the ground. Now shoot the cannonball again, this time with a higher forward speed. The ball still falls and eventually strikes the earth, but because it has a higher forward speed (sideways, relative to the earth) the ball can cover more distance before striking the earth. If you shoot the ball fast enough, it will still fall but will never manage to strike the earth. The earth will curve away faster than the ball can fall towards it. As a result, the ball will continually fall and miss and will end up circling the earth.

All objects in the universe are constantly falling.

Newton cannonball thought experiment
Newton's cannonball thought experiment demonstrating that orbits are just objects falling but travelling sideways fast enough to never strike the source of the gravity.
Credits: Source - Christopher S. Baird.
Date: July 1, 2013
https://www.wtamu.edu/~cbaird/sq/2013/07/01/why-doesnt-the-earth-fall-down/
Reproduced with permission

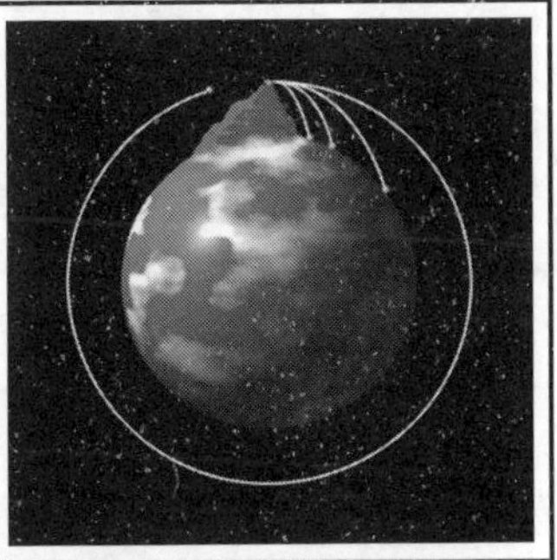

Note: Centrifugal Force: An apparent force that is felt by an object moving in a curved path that acts outwardly away from the centre of rotation. **Centripetal force:** A force that acts on a body moving in a curved path and is directed towards the centre around which the body is moving.

(2). The other way is according to Einstein's general relativity

In this way of looking, the moon is not falling towards the earth, it is falling sideways around the earth. Einstein's Theory of General Relativity explains the properties of space and time in the universe. Within that theory, objects with mass curve spacetime in their vicinity, and this curvature influences the motions of other objects. The greater the mass and density of the object, the larger the curvature of spacetime that results. So, the moon orbits the earth because the earth curves spacetime in the vicinity of the moon. Though the moon itself curves space as well (since it has mass), the curvature in the vicinity of the moon is dominated by the earth, which tells the moon to orbit the earth, given the moon's current position and motion in the sky. It is this interplay between mass and curvature that causes the gravitational force which is a centripetal force.

An artist concept of an object orbiting the Earth (Note: in the image Gravity Probe is shown as an object orbiting earth. Gravity Probe was a satellite-based experiment which confirmed predictions of general relativity)
Credits: Author NASA. May 18, 2008
Source: http://www.nasa.gov/mission_pages/gpb/gpb_012.html
Image is in public domain

References

1. *Baird, Christopher S. Why does't the earth fall down. West Texas A&M University website "Science Questions with Surprising Answers". July 1, 2013*
   https://www.wtamu.edu/~cbaird/sq/2013/07/01/why-doesnt-the-earth-fall-down/
2. *Spekkens, Kristine. How does the moon stay "suspended" in the air? (Intermediate). Curious About Astronomy? Ask an Astronomer (run by volunteers in the Astronomy Department at Cornell University). Last updated on July 18, 2015.*
   http://curious.astro.cornell.edu/about-us/44-our-solar-system/the-moon/general-questions/109-how-does-the-moon-stay-suspended-in-the-air-intermediate
3. *Walker, Robert. Why does the Earth not fall into the sun or the moon fall into the earth? Quora. Updated May 5, 2021*
   https://www.quora.com/Gravity-Why-doesnt-the-earth-fall-into-the-sun-or-the-moon-fall-into-the-earth

Moon's Rotational Motion

Q138. How long is a day on the moon?

A138. The rotation period of the moon around its axis is 27.322 days. So, a typical "moon Day" is 13.661 earth days, so is the night.

Q139. Why do we see only one side of the moon?

A139. We only see one side of the moon, known as "Near Side" (much on this later). It is because the moon's rotation period is exactly equal to the time it takes to orbit around the earth, 27.322 earth days (effectively, moon's day is as long as its year). This is known as synchronous rotation. If the moon didn't rotate at all, we would see all its sides.

The underlying reason of synchronous rotation is because the moon has been tidally locked to the earth. Tidal locking occurs when the gravitational interaction of two astronomical bodies drive each other towards a synchronous rotation. Let us look at the earth-moon system. The moon's tidal forces causes the high and low tides on the earth. In the same way, the earth exerts tidal forces on the moon. These tidal forces exert torques on the rotating system and tends to slow its rotation till it finally shows the same face towards the other body. So, it is the effect of tidal forces of the earth on the moon that have caused the moon to show only one face to the earth. In the beginning, the moon was rapidly spinning close to the earth. Then earth's gravity began to take effect. Tidal forces drifted the moon away from the earth in its orbit and slowed its spin. Eventually, the moon reached a point where one side was always facing the earth. The moon still drifts further away a few centimeters a year. The moon and earth's gravitational tug of war is slowing down the earth's rotation a tiny bit too. This would continue until both the earth and moon were 'tidally locked'. In that situation, both the moon and earth would spin at the same rate of 47 days, showing the same face to each other for the rest of their existence. Such an example is the dwarf planet Pluto and its largest moon Charon which are both tidally locked to each other.

Q140. What are the sidereal and synodic months and why are they different?

A140. Sidereal Month is the time the moon takes to complete one orbit around the earth with respect to the background stars. It is 27.322 days.

The moon's Synodic Period is the amount of time it takes for the sun to return to the same position in the sky, that is, same moon phase, say "New moon", which is 29.531 days. This is the month used in lunar calendars and is also known as lunar month.

You may ask why a synodic month is longer than a sidereal month.

The answer is that since the earth is constantly moving along its orbit about the sun, the moon must travel slightly more than 360° to get from one new moon to the next. Thus, the synodic month is longer than the sidereal month. To understand it better, consider a "New moon Phase" as day zero. sun-moon-earth are aligned. For a "Lunar sidereal month", It will take the moon 27.322 days to complete one orbit around the earth. During this period the earth has moved ahead along its orbit. But as you can see from the figure, the next "new moon" position has not arrived yet. The moon and the earth must move together for an extra 2.2 days to re-align with the sun to get the next new moon phase.

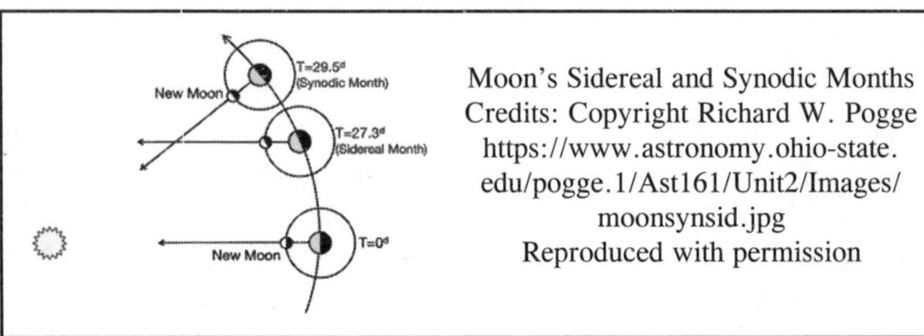

Moon's Sidereal and Synodic Months
Credits: Copyright Richard W. Pogge
https://www.astronomy.ohio-state.edu/pogge.1/Ast161/Unit2/Images/moonsynsid.jpg
Reproduced with permission

Q141. Why does the moon rise and set at different times and in different locations?

A141. Just like the sun and the nighttime stars, the moon rises in the east and sets in the west each day due to earth spinning, and not from the moon's orbit around the earth.

The lunar movement can be spotted by looking at where the moon is in relation to stars in the background one night, and then comparing to where it is several hours later or on the next night. You will notice it has moved east. This movement is from the moon's orbit, which takes 27 days, 7 hours, and 43 minutes to go full circle. It causes the moon to move 12–13 degrees east every day. This shift means earth must rotate a little longer to bring the moon into view (moon rise), which is why moonrise is about 50 minutes later each day. As it rises later, the moon appears in a different part of the sky.

Reference

Rothstein, Dave. Why is the moon in a different place every night? (Beginner). Curious About Astronomy? Ask an Astronomer (run by volunteers in the Astronomy Department at Cornell University). last updated on July 18, 2015. http://curious.astro.cornell.edu/observational-astronomy/46-our-solar-system/the-moon/observing-the-moon/141-why-is-the-moon-in-a-different-place-every-night-beginner

Q142. What is the Earth – Moon Tango?

A142. The earth goes around the sun, and the moon moves around the earth. That's an earth-moon tango around the sun.

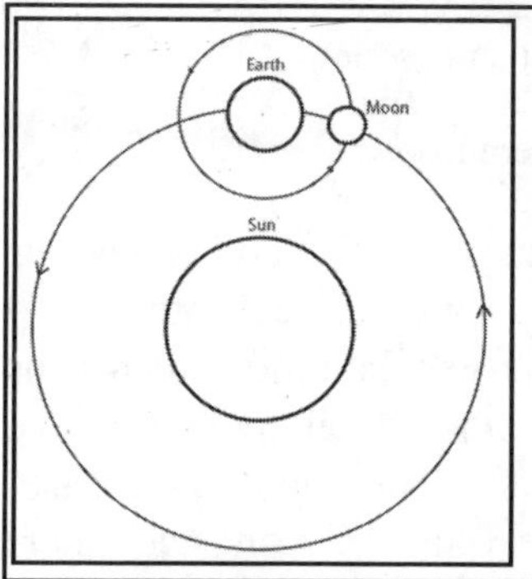

Sun, Earth, and Moon Model (Not to scale)
Drawn by Mahika (Mahi) Inaganti
Reproduced with the permission

If we join the earth and the moon by an invisible and a featherweight rod, like a dumbbell of sort, the whole thing will balance at a point of their centre of mass. This is the barycentre of the earth-moon system. The barycentre does not lie exactly in the middle of the two. It sits just inside the earth, about 1,700 kilometres (1,060 miles) inside along the earth-moon line. Therefore, it is more accurate to say that it is the barycentre, rather than the centre of earth, that follows an elliptical path around the sun. An animated movie of the earth and the moon around the sun looks quite like a tango.

The orbital geometry of the moon, earth, and the sun gives rise to the moon's phases and to the phenomena of lunar and solar eclipses (more on these later).

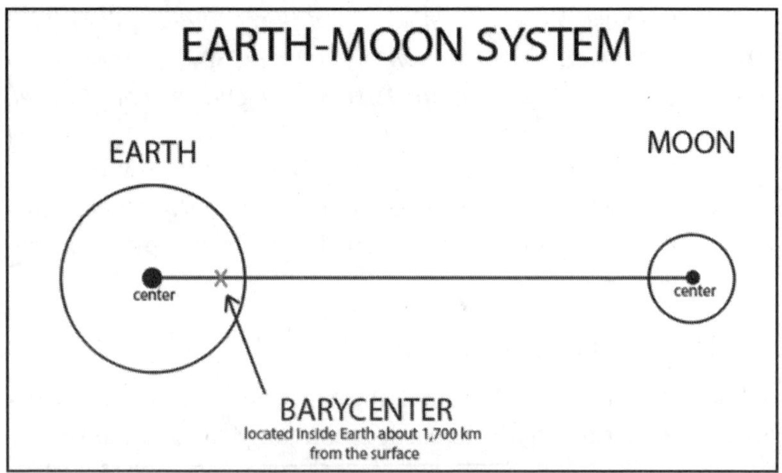

Earth – Moon Barycentre (not to scale)
Drawn by Mahika (Mahi) Inaganti
Reproduced with the permission

The earth-moon system has distinct properties and behaviour of its own. This is because of their nearness to each other, and the ratio of their masses being much larger than those of moons of the planets that they orbit (the earth moon is relatively massive compared to moons of other planets). The moon and earth consequently exert a strong gravitational influence on each other.

The motion of the moon is really a 3-body process, influenced by the sun as well, with further perturbations perhaps due to Jupiter etc. The orbit is close to a circle (low eccentricity), which means that a pull of a few 1000 kilometres this way or that can shift the time of largest and smallest distance by a great amount, in contrast to what it would do to a high-eccentricity orbit. The distance between the moon and earth varies widely because of the combined gravity of earth, the sun, and the planets. For example, in the last three decades of the twentieth century, the moon's apogee—the farthest distance that it travels from earth in a revolution—ranged between about 404,000 and 406,700 km (251,000 and 252,700 miles), while its perigee—the closest that it comes to earth—ranged between about 356,500 and 370,400 km (221,500 and 230,200 miles).

Q143. Why does the moon wobble?

A143. The moon "Wobble" from left to right. Like all celestial objects, with elongated elliptical orbits, moon's speed varies on its path around the earth following the Kepler's laws of motion. The moon speeds up when it is at perigee (closer to earth) and slows down when it is at its apogee (farther from earth). That's because earth's gravity pulls harder on the moon the closer it is. The moon's rotational speed is constant. So, at perigee moon's orbital speed is faster than its rotational speed. This causes a slow rocking back and forth known as "lunar libration". This allows an observer on earth to see slightly more of moon's surfaces at the eastern limb (edge) or, at the western limb at different times. As a result of this, about 59 per cent of moon's surface becomes visible from the earth over its entire monthly cycle. However, at any given moment, one can only see 50 per cent.

Q144. What is moon's apsidal/orbital precession?

A144. Moon's orbit is not fixed in space. If you join the perigee and the apogee to produce a line, it is called the line of apsides. This line continuously keeps changing its direction in space. As a result, the moon's orbit as a whole rotates slowly. It makes one complete rotation in about 8.85 earth years. This rotation is known as apsidal or orbital or precession.

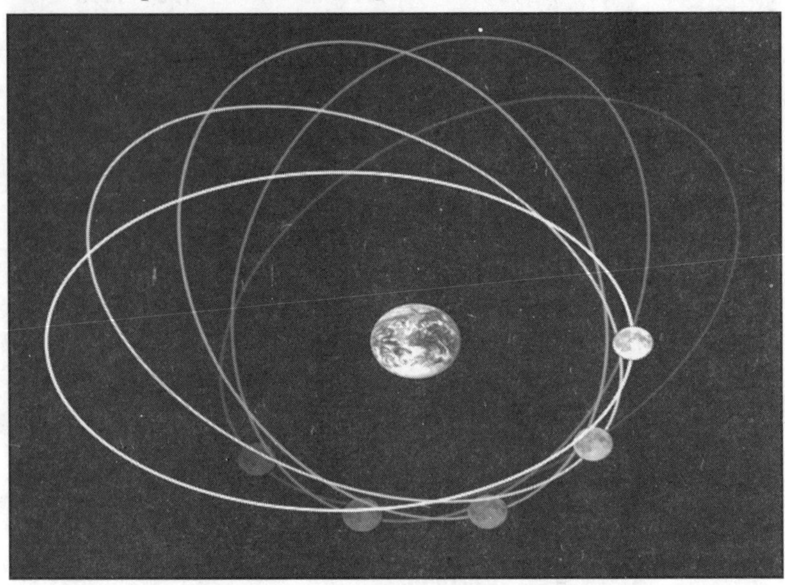

Moon's Apsidal/Orbital Precession
Credits: Wikimedia Commons,
Author: Rfassbind, August 22, 2015
https://en.wikipedia.org/wiki/Lunar_precession#/media/File:moon_apsidal_precession.png
Figure in Public Domain

❑

CHAPTER 12

Near Side and Far Side of the Moon

Q145. What are near side and far side of the moon?

A145. As the moon orbits around the earth, it always keeps the same face towards earth and the other side perpetually keeps facing away. So, the moon has a near side, the lunar hemisphere, that always faces towards the earth and a far side, the other hemisphere, which remains hidden from it. This happens because the moon rotates on its axis at the same rate that the moon orbits the earth–a situation known as tidal locking. However, it was not like this in the beginning. The moon was spinning at a much faster pace. Under the earth's gravitational influence, the moon's rotation has gradually slowed down to match its revolution speed. Now the moon's rotation period is locked into its revolution period. This is called synchronous motion. (More on this later)

Q146. The far side of the moon is also referred to as "dark side". Does it mean it never receives sunlight?

A146. Many people incorrectly believe that the moon's far side never sees light of the day and they often refer to it as the dark side. This is a misconception. In fact, the moon is ceaselessly rotating on its axis, and no part of the moon permanently remains engulfed in darkness (except some deep craters in the polar regions). During a full moon phase, the near side receives direct sunlight, and the far side remains dark. On the other hand, during a new moon phase, the near side becomes dark, and the entire far side becomes sunlit. This is day and night of the moon. So, both sides receive direct sunlight for the same length of time. Hence, the phrase "dark side" rather figuratively refers to the unknown or unseen side.

Q147. Have we seen/visited the far side of the moon?

A147. Soviet Union's Luna 3 photographed moon's far side for the first time in 1959. The astronauts of Apollo 8 were the first men to see the far side of the moon in with their own eyes in 1968 and the astronauts were able to describe the surface of the moon in detail. Astronaut William Anders reported: "The backside (of the moon) looks like a sand pile my children have played in for some time. It's all beaten up, no definition, just many bumps and holes". Though, a dozen astronauts have landed on the moon's surface so far, it's far side is yet to receive one. The China National Space Administration's (CNSA) Chang'e 4 spacecraft made the first ever soft landing on the lunar far side on January 3, 2019 and deployed Yutu-2 lunar rover onto far side lunar surface.

Q148. What are major differences between near side and far side of the moon?

A148. Scientists have yet to fully understand the differences between the near side and the far side of the moon. This mystery, called the "Lunar Farside Highlands Problem" by astronomers, dates to 1959, when the Soviet spacecraft Luna 3 transmitted the first images of the far side of the moon back to earth. Researchers immediately noticed that fewer maria on the far side of the moon.

Some of the major differences between the two sides are:

1. The near side of the moon is characterized by its large, dark, lava-filled, plain areas called maria. The far side's terrain is rough and rugged. There is a multitude of impact craters.

Only one per cent of the far side is covered with maria compared with approximately 31 per cent for the near side.

2. The near side is low and flat while the far side is mountainous.
3. The crust on the farside is about 50 kilometres thicker than that on the nearside (average). That explains the lack of maria on the far side. The thicker crust means it was harder for giant impacts to pierce the crust and get darker basaltic lava bubbling up from below.
4. The far side is shielded from radio transmissions from the earth. It is, therefore, considered a good location for placing radio telescopes.
5. The far side is expected to have the highest concentration of helium-3 because the near side is partly shielded from the solar wind by the earth. Helium 3 is relatively rare on earth but has good potential for use as a fuel in nuclear fusion reactors (more on this later).
6. The nearside is richer in potassium (K), rare-earth elements (REE) and phosphorus (P), components collectively known as KREEP. Crust-forming models show that these would have been concentrated in the last remnants of subsurface magma to crystallize as the moon cooled.

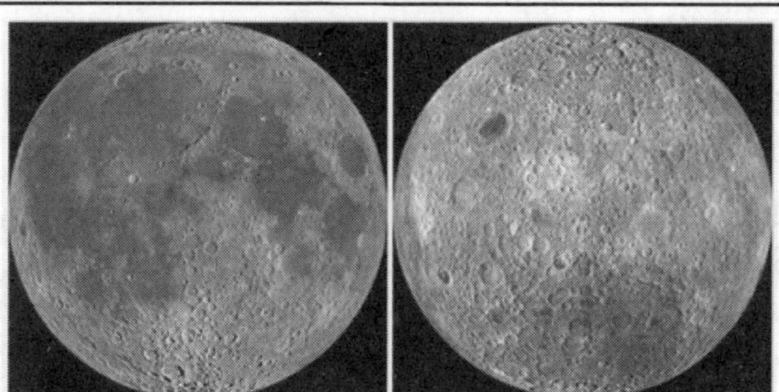

The moon's two sides: Near side on left and the far side on right.
The near side contains 97% of the dark volcanic plains, while the far side is much more cratered
Composite images taken by Lunar Reconnaissance Orbiter (LRO) in June 2009
Credits: NASA/GSFC/Arizona State University

Q149. Why are the two sides of the moon (near side and far side) different from each other?

A149. Scientists are searching for answers to puzzling questions about the moon's two sides, such as why is the far side more cratered than near side? Why is the near side so much smoother, and dominated by vast basaltic, volcanic plains? Why was this ancient volcanic activity mostly restricted to the near side?

With the help of mathematical models of the early earth-moon system and other scientific studies, scientists have proposed the following ideas to explain some of the differences.

1. Two Moons Theory or the second impact theory

Scientists were puzzled by the fact that all planets that have moons have more than one moon except the earth which has only one. In 2011 "two moons" theory was announced by planetary scientists Erik Asphaug and Martin Jutzi. The giant impact, which formed the moon, simultaneously created one more moon. After some time two moons collided and coalesced. This was a low-speed collision and less explosive. The smaller moon splashed on the far side and flowed like liquid over it. This created an extra thicker crust on the far side. NASA tried to test this theory, but the results were inconclusive. While there was no evidence supporting the theory, the NASA mission did not disprove it either.

2. Bright and hot earth facing moon's near side shortly after the giant impact that formed the moon

The second theory was announced in 2014 by Jason Wright, Steinn Sigurdsson, and Arpita Roy, all Penn State astrophysicists as indicated by A'ndrea Elyse Messer. The researchers believe that the absence of maria on the far side of the moon, which is due to a difference in crustal thickness between the near side of the moon and the far side, is a consequence of how the moon originally formed.

Shortly after the giant impact that formed the moon, the earth and the moon were very hot. The moon was 10 to 20 times closer to earth than it is now, and it quickly assumed a tidally locked position with the earth. Because the earth and the moon were tidally locked from the beginning, the still hot earth - more than 2500 degrees Celsius (4500 F) - radiated towards the near side of the moon. The far side, away from the boiling earth, slowly cooled, while the earth-facing side was kept molten creating a temperature gradient between the two halves. Earlier in its history, large meteoroids struck the nearside of the moon and punched through the crust, releasing the vast lakes of basaltic lava that formed the nearside maria. Meanwhile, when meteoroids struck the far-side of the moon, in most cases the crust was too thick and no magmatic basalt welled up, creating the far side of the moon with valleys, craters and highlands, but almost no maria.

So, it was the young earth glowing bright and hot in the moon's sky that made the two sides so different.

3. Differences in geochemical composition between the two sides

A new study (year 2020) by Elardo and others at the Florida University believe that the relative darkness of the patches in the near side of the moon was due to their geological composition attributable to volcanism. Scientists have identified a new type of rock associated with the maria terrane of the near side named KREEP unlike anywhere else on the moon. KREEP is short for rock enriched in potassium (chemical symbol K), rare-earth elements (REE, which include cerium, dysprosium, erbium, europium, and other elements which are rare on earth) and phosphorus (chemical symbol P). The KREEP terrane also contains the elements Thorium (Th) and Uranium (U), which decay radioactively and produce heat. The heat from the radioactive decay of these elements can help melt the rocks they are contained in. The scientists studied the effect of KREEP in volcanic activity. Their study showed that compositional asymmetries between both the sides began to have a dramatic effect on magma production immediately after lunar differentiation. The large concentration of heat-producing elements on the moon's nearside acted as a heat source for melting and lowering melting temperatures at the crust–mantle interface. This produced more crust-building magmas than would have occurred on the farside. The results from this study suggest that the moon's KREEP-enriched maria have influenced lunar evolution since the moon formed.

References

1. *Lovett, R. Early Earth may have had two moons. Nature. August 3, 2011. https://doi.org/10.1038/news.2011.456* https://www.nature.com/articles/news.2011.456
2. *Jutzi, M and Asphaug, E. Forming the lunar farside highlands by accretion of a companion moon. Nature 476, 69-72. August 3, 2011.* https://pubmed.ncbi.nlm.nih.gov/21814278/
3. *Messer, A'ndrea Elyse. 55-year-old dark side of the moon mystery solved. Penn State News. June 9, 2014.* https://news.psu.edu/story/317841/2014/06/09/research/55-year-old-dark-side-moon-mystery-solved
4. *Roy, Arpita et al. Earthshine on a Young Moon: Explaining the Lunar Farside Highlands. The Astrophysical Journal Letters, Volume 788, Number 2. June 20, 2014* https://iopscience.iop.org/article/10.1088/2041-8205/788/2/L42
5. *Gough, Evan. Do We Now Understand Why the Moon's Near and Far Sides Look So Dramatically Different? Universe Today. June 26, 2020.* https://www.universetoday.com/146713/do-we-now-understand-why-the-moons-near-and-far-sides-look-so-dramatically-different/
6. *Why the far side of the moon so different from near side (Scientists find that these are linked to an important property of rock signature KREEP). The Tribune India. Jun 22, 2020* https://www.tribuneindia.com/news/schools/why-the-far-side-of-the-moon-so-different-from-near-side-102715
7. *Elardo, S.M., Laneuville, M., McCubbin, F.M. et al. Early crust building enhanced on the Moon's nearside by*

mantle melting-point depression. Nat. Geosci. 13, 339–343 (2020). Published March 30, 2020
https://doi.org/10.1038/s41561-020-0559-4
https://www.nature.com/articles/s41561-020-0559-4#citeas

Q150. How can the far side of the moon be beneficial to glimpse the beginnings of the universe?

A150. Each object in the cosmos gives off unique patterns of radio emissions. This allows astronomers to get the whole picture of a distant object. We also need these signals to learn how the universe was formed (inflated) immediately after the Big Bang.

Earth-bound radio telescopes encounter too much interference from electromagnetic pollution caused by human activity such as cell phone towers, Wi-Fi hotspots, and numerous devices giving off radio waves. earth's ionosphere also blocks radio waves.

The far side of the moon is the perfect place for radio astronomy where we will not get disturbed by the earth. It is one of the best places in the inner Solar System for placing radio telescopes to monitor low-frequency radio waves. A team of astronomers recently recommended to set up a sensitive SETI instrument on the moon's far side and using it to scan space.

Note: The "Search for Extra-Terrestrial Intelligence" (SETI) is a collective term for scientific searches for intelligent extra-terrestrial life, for example, monitoring electromagnetic radiation for signs of transmissions from civilizations on other planets

References

1. *Williams, Matt. The Moon is the Perfect Spot for SETI. Universe Today. October 11, 2020.*
https://www.universetoday.com/148146/the-moon-is-the-perfect-spot-for-seti/
2. *Silk, Joseph. Put telescopes on the far side of the Moon. Nature. January 3, 2018*
https://www.nature.com/articles/d41586-017-08941-8

CHAPTER 13

Moon Phases

Q151. What causes moonlight?

A151. The ancients thought that the moon produced its own light, but we now know that this is not true. The moon doesn't produce its own light but shines because it reflects light from the sun. It was Anaxagoras (500 BC–428 BC), a Greek philosopher, who first explained that the moon shines due to reflected light from the sun. He attempted to give a scientific account of eclipses, meteors, rainbows, and the sun. Anaxagoras was imprisoned for claiming that the sun was not a god and that the moon borrowed its light from the sun.

Q152. Why is the moon so bright?

A152. The full moon is by far the brightest object in the night sky.

The brightness of a planet or a moon depends on three things:

(1) The object's albedo: Albedo is the proportion of the incident light or radiation that is reflected by a surface, typically that of a planet or moon.

(2) The total amount of light that is hitting the object.

(3) The distance between the object and the eye or camera that is viewing it

The strongest source of light in our solar system is the sun. We see planets and moons because they reflect sunlight. The amount of sunlight incident on a moon or planet that gets reflected depends on:

1. The materials on its surface.
2. Its atmosphere.
3. Its surface roughness.

Snow, rough ice, and clouds are highly reflective, most types of rocks are not. Therefore, a planet that is covered with clouds, such as earth or Venus, is generally brighter than a rocky moon or planet.

The moon is one of the dimmest objects in our solar system compared to other astronomical bodies. It is because the moon is the least reflective object in the solar system. It reflects only about 12 per cent (albedo 0.12) of the sunlight that hits it (about the same as old asphalt, such as found in roads and parking lots). It only seems bright in the night sky because it is so close to the earth and because the trees, houses, and fields around us are so dark at night. The next brightest object in the sky is Venus. The moon appears about 1,500 times brighter than the Venus. Sirius is the brightest star in the sky. The moon appears some 27,000 brighter than Sirius. The other way to put it, the moon shines more than 40 times brighter than all the other objects in the night sky combined. And it is all because the moon is very near to earth.

Distance between earth and the moon 384,400 km (238,900 mi), Albedo 0.12

Distance between earth and Venus about 40 million kilometres (25 million miles), Albedo 0.77

Distance between earth and Sirius star 8.611 light years

References

1. *Baird, Christopher S. Why is the moon so bright? Science Questions with Surprising Answers. August 6, 2015.* https://www.wtamu.edu/~cbaird/sq/2015/08/06/why-is-the-moon-so-bright/
2. *Siegel, Ethan. Ask Ethan: How Bright Is The earth As Seen From The Moon? Starts With a Bang. Forbes. March 18, 2017.* https://www.forbes.com/sites/startswithabang/2017/03/18/ask-ethan-how-bright-is-the-earth-as-seen-from-the-moon/?sh=13b5cdf1769d

Q153. Is there any difference in appearance of the moon viewed in southern hemisphere (say Australia) and northern hemisphere (say New York)?

A153. To Australians and other people living in places in the southern hemisphere the moon appears "upside down" (reversed top to bottom) compared to northern hemisphere.

The moon seen from its orbital plane
https://www.planetary.org/space-images/the-moon-seen-from-its-orbital-plane
Credits: The Planetary Society
Reproduced with permission

It is because our earth is spherical. Imagine it this way: if you stand at the north pole, with your head "up," and your friend stand on the south pole, with his head "up," relative to the ground, the two heads are pointed in exactly opposite directions. If you both look at the moon, then your friend in the southern hemisphere will see the moon's dark 'Oceanus Procellarum' mare in the south-east corner compared to what you, in the northern hemisphere, see it in the north-west corner.

It is not only the moon, other heavenly bodies, like constellations, do the same thing.

So, the moon is not upside down, but our orientation changes depending on if you're north or south of the equator. Physically, we are upside down compared to someone standing in the opposite hemisphere.

> **Note:** Oceanus Procellarum is a vast lunar mare on the near side of the moon. It is the largest of the maria ("seas"), stretching more than 2,500 km (1,600 mi) across its north–south axis and covering roughly 4,000,000 sq km (1,500,000 sq mi), accounting for 10.5 per cent of the total lunar surface area.

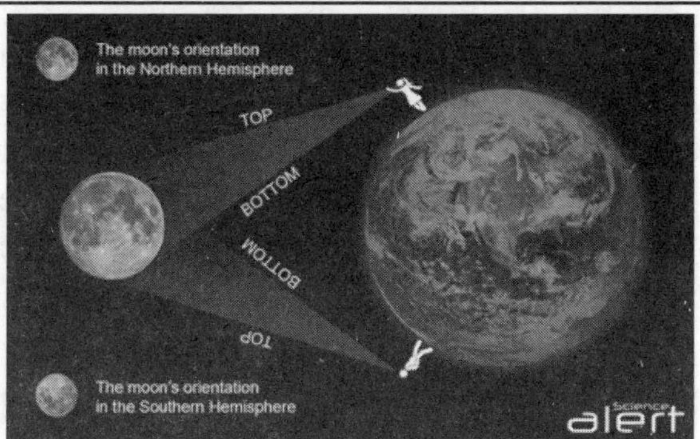

The moon and the night sky is actually rotated 180 degrees as viewed from the Northern and Southern hemisphere.
Credit: Tessa Koumoundouros and ScienceAlert.
https://kottke.org/18/11/the-moon-is-flipped-upside-down-in-the-southern-hemisphere
Reproduced with the permission

References

1. *Kottke, Jasion. The Moon is Flipped Upside Down in the Southern Hemisphere. Kottke.org. November 21, 2018* https://kottke.org/18/11/the-moon-is-flipped-upside-down-in-the-southern-hemisphere
2. *Blog by Science fiction author C. Stuart Hardwick published in Quora answering a question: "Do people in the most Southern Hemisphere see the other side of the moon?" updated April 24, 2018* https://www.quora.com/Do-people-in-the-most-Southern-Hemisphere-see-the-other-side-of-the-moon

Q154. Why does the moon seem to change its shape every night?

A154. The changing shapes of the moon that are seen from the earth are known as "phases" of the moon. There is a misconception about moon phases that these are caused by a shadow of the earth (That is eclipse not phase). The moon phases are due to its orbit around the earth.

In our entire solar system, the only object that shines with its own light is the sun. The moon, earth and other celestial objects reflect sunlight. We see the moon because sunlight gets reflected to us from the moon's surface. Light from the sun hits the moon and always lights up half of the moon because the moon is spherical.

The lunar phases are created by changing angles (relative positions) of the earth, the moon, and the sun, as the moon orbits the earth (remember, the moon is orbiting earth and earth orbiting sun). As the moon moves, we see its different fractions of the illuminated half. For example, When the moon and the sun are on opposite sides of the earth, we perceive the moon as full, known as "full-moon". However, when the sun and the moon are on the same side of the earth, we do not see the moon and we say it as the "new-moon". During a new moon, the side of the moon that we can see from earth is not illuminated by sunlight at all, the other side is illuminated.

In other words, in each phase, only a portion of the illuminated half of the moon is visible to earth, ranging from 0 per cent at a new-moon to 100 per cent during a full- moon

It takes about 29.5 days for the moon to make a complete trip around earth. Therefore, each phase repeats itself every 29.5 days.

Mercury, Venus, and Mars also have phases like the moon which can be spotted at certain times with a telescope. And the earth has phases too. When Apollo astronauts looked back on earth from the moon, they didn't just see a globe, they saw different phases of earth depending on which phase the earth was going through.

You can understand moon-phases by a simple experiment. Hold a ball in your hand. The ball represents the moon and you are earth. Now ask someone to hold a flashlight several feet away from you. Flash represents sun. Flashlight will shine the "moon" ball (flashlight is getting reflected by the ball, that is why you see the ball). Now slowly turn around keeping the ball in your hand, watching the moon go around you. You will see the ball moon phases.

Q155. How were the moon's phases named?

A155. Each day the moon moves leftward (east) through the sky – for people looking at it from the northern hemisphere. As it moves through a 29.5-day orbit, the moon starts New, grows Full, then wanes to new again.

The moon phases are named using the words describing the "age" of the moon, the apparent "shape" of the moon, its direction of growth, and its location in its orbit around the earth.

Let us understand a few terms:

"New" and "Full": When the moon is not readily visible at all, it's a "new moon"; when the whole surface facing the earth is lighted, it's a "full moon."

"Waxing" and "Waning": As the surface of the moon becomes more lighted each day on the way to full, it is "waxing." "To wax" means to increase in size gradually. And as the surface becomes less lighted each day on the way back to "new," it is "waning," which, means to decrease in size gradually.

"Crescent" and "Gibbous": Just before and after "new" moon, we have a "crescent moon," so called because only a small arc is visible. Then there's the "gibbous moon" (gibbous in Latin means hump) when it is more than a semi-circle. So, we can have a "waxing gibbous moon" and a "waning gibbous moon,"

Here are the 8 moon phases in order:

The new moon, first quarter moon, full moon, and the third quarter moon (also known as last quarter moon), are moon's four principal phases. During the intervals between principal phases, the moon appears as crescent or gibbous. These are called the intermediate phases. The moon is crescent when it is less than half illuminated and gibbous when it is more than half illuminated. Between a new moon and a full moon, the moon is growing or waxing phase. This is also called "shukl paksha" in India. The moon is in a shrinking or waning phase between a full moon and a new moon. This is "krishna paksha".

The moon's eight principal and intermediate phases are given the following names, in sequential order.

1. **New Moon** – Also known as the "Dark moon" as the moon's near side faces away from the sun and it is nighttime there. This happens when the moon lies between the sun and the earth in an approximate alignment. Moon's entire sunlit part is on the far side facing away from the earth.

2. **Waxing Crescent Moon** – During this period the moon's near side reflects just a sliver of light from the sun and it grows day by day starting from a new moon. This is called "Shukla Paksha" in India.
3. **First Quarter Moon** – Commonly called the half-moon, as sun, earth and the moon form a 90-degree angle and you see about half of moon's sunlit surface. You will see the moon shining bright at sunset in the middle of the sky.
4. **Waxing Gibbous Moon** – During this period the sunlit part of moon's disc grows towards full.
5. **Full Moon** – moon's entire near side glows full, lit by the sun. In this phase, the moon is again in approximate alignment with the earth and the sun, but this time the earth lies between the sun and the moon as the moon has moved to the opposite side in orbit around the earth, and moon's entire shadowed back side faces away. As the moon is directly opposite the sun, it rises at sunset, stays visible all night long and sets just as the sun rises the next morning.
6. **Waning Gibbous Moon** – The moon is more than half full but on the wane. It's "Krishna Paksha" starting after the full moon. The moon starts rising after sunset late every day by about an hour.
7. **Last Quarter Moon** – This is another half-moon. The new moon is still about a week away. The sun-earth-moon angle is 90 degrees again. This time, however, you will see the moon at sunrise in the middle of the sky.
8. **Waning Crescent Moon** – Just about a sliver of the moon's near side glows bright a few days before the next new moon.

Each of the four "intermediate" phases is roughly seven earth days (~7.4 days) long but, this varies slightly due to moon's elliptical orbit.

Some interesting points about the names of Moon's phases:

1. The new moon is the commonly accepted "beginning" and also the "end" of the cycle.

2. "Wax" and "Wane" are old words, originally from Sanskrit that made their way through history to old German and finally to old English.

Phases of the Moon
This graphic shows all eight moon phases. It shows the position of the moon and the sun during each of the moon's phases and the moon as it appears from earth during each phase. The moon makes a complete orbit of earth about every four weeks.
Not to scale.
Credit: Copyright Richard W. Pogge.
https://www.astronomy.ohio-state.edu/pogge.1/Ast161/Unit2/Images/moonphase.gif
Reproduced with Permission

3. Astronomers use the "quarters" to talk about the orbit of the moon and its location in the orbital path. The name "quarter" says "the moon is a quarter of the way through its orbit now." In naming the moon, it has a "First Quarter" and a "Third Quarter", but no "Second Quarter" and no "Fourth Quarter" or "Zeroth Quarter." The Second Quarter moon is named as full-moon and the fourth quarter (also the zeroth quarter) is named as new-moon.

4. The moon looks like a "half-moon" two times during the moon's cycle. It is a half-moon as it grows (waxes) and becomes

a half-moon again when it shrinks (wanes). Confusingly, astronomers call the "half-moon" a "quarter-moon." (First quarter-moon and third-quarter moon). Daniel Cummings, an astronomer (https://starinastar.com) has rightly suggested that we should call the first quarter- moon the "waxing half-moon" and the third quarter moon the "waning half-moon".

References

1. *Moon in Motion. NASA – Earth's Moon*
https://moon.nasa.gov/moon-in-motion/moon-phases/
This site is maintained by the Public Engagement Team at NASA Goddard's Solar System Exploration Division for the Science Mission Directorate.
NASA Official: Lori Glaze
Producer: Molly Wasser
Content Development: Andrea Jones, Caela Barry, Tracy Vogel
Graphics: Vi Nguyen
2. *Cummings, Daniel. Moon Phases Names – an easy way to remember. Star in a Star (Learning Astronomy, Teaching Astronomy). March 6, 2019.*
https://starinastar.com/moon-phases-names-an-easy-way-to-remember/

Q156. Are moon phases the same everywhere on earth? In other words, is it the full-moon in Sydney at the same time as the full-moon in Delhi, London, and New York?

A156. The answer is Yes.

Regardless of where we are on earth, when it is a full moon, it is a full moon everywhere. Same for new moons and all the other phases. Everyone sees the same phases of the moon, but people south of the equator will see the moon upside down so that the reverse side is lit. When the moon is "C" shape in Northern hemisphere, it will appear as "D" shape in southern hemisphere.

But keep in mind that the clock times will be different at different geographical locations based on global time-zones. In other words: When we say that there is a "full-moon coming up in three hours" – that is the case around the world. If it is coming up in three hours in Sydney, it is also coming up in three hours in Delhi, London, New York, San Francisco, or wherever we are. The clock times, of course, will be different at all these places.

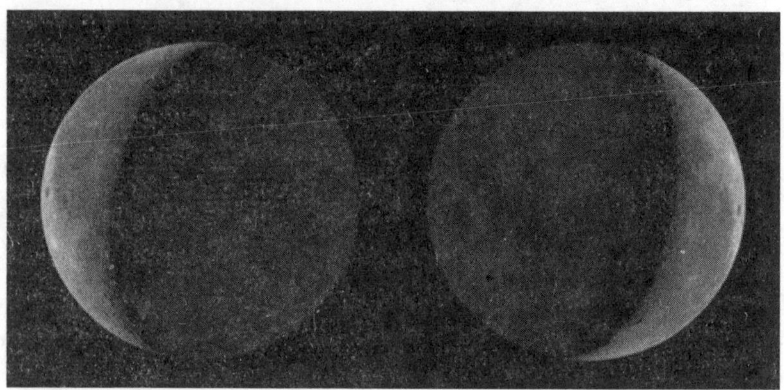

Data visualizations of the waning crescent moon as seen from the Northern Hemisphere (left) and the Southern Hemisphere (right) on the same day. Credit: NASA's Scientific Visualization Studio

Q157. What are the many names given to the full moons?

A157. Each full moon of the year has its own name.

'Full moons come, Full moons go,
Softening nights, with their silver grow,
They pass in silence all untamed
But as they travel, they are named '

Rhyme from "When the Moon is full" a book for young children by Penny Pollack (pub' Little Brown)

The movement of the moon is used to keep track of time and planning for activities like hunting, planting, and harvesting. Ancient cultures, the world over, have given these full Moons names based on the plants, animals, or weather during that month. The Maine Farmers' Almanac first published Native American names for the full Moons in the 1930s, and over time these names have become widely known and used. The following is a list of some of the most common names for the full moon in the Northern Hemisphere, with alternate names in parentheses.

January : Wolf Moon (Old Moon, Moon After Yule)

February : Snow Moon (Hunger Moon)

March : Worm Moon (Crow Moon, Sap Moon)

April : Pink Moon (Egg Moon, Grass Moon)

May : Flower Moon (Milk Moon, Planting Moon)

June : Strawberry Moon (Flower Moon, Rose Moon)

July : Buck Moon

August : Sturgeon Moon (Grain Moon, Green Corn Moon)

September : Harvest Moon (Fruit Moon)

October : Hunter's Moon

November : Beaver Moon (Frosty Moon)

December : Cold Moon (Full Long Nights Moon)

Hindu full Moon festivals

Purnima is the Indian and Nepali word for full moon. In the Hindu religion, each Purnima has its own festival name and seasonal traditions, in tune with planting, harvest, return of longer days. Full moon day is considered auspicious for new beginnings. Many of the Hindu festivals are celebrated on full moon days. Different parts of India celebrate the same day with different names, as listed below:

1. **Chaitra Purnima** - Gudi Padua, Yugadi, Ugadi, Hanuman Jayanti
2. **Vaishakh Purnima** - Narasimh Jayanti, Buddha Jayanthi
3. **Jyeshtha Purnima** - Vat Savitri Vrat Vat Purnima
4. **Ashadh Purnima** - Guru Purnima
5. **Shravan Purnima** - good day for starting Upanayan day, Avani Avittam, Raksha Bandhan- conceptually Onam also comes on this day.

6. **Bhadrapad Purnima** - start of Pitrupaksha, Madhu Purnima
7. **Ashvin Purnima** - Sharad Purnima
8. **Kartik Poornima** - Thrukkarthika
9. **Margasirsha Purnima** - Thiruvathira, Dathatreya Jayanthi
10. **Pushya Purnima** - Thaipusam, Shakambharee Purnima
11. **Magha Purnima** – Vishnu Poojan, Megh Mela
12. **Phalguna Purnima** - Holi

In Sri Lanka, the full moon corresponds with the Nikini Poya holiday, commemorating the first Buddhist council, which occurred about 2,400 years ago, sometime around 400 BCE. In Kandy, a large city in central Sri Lanka, the full moon corresponds with the end of the Esala Perahera festival, also known as the Festival of the Tooth, a two-week Buddhist festival held each year.

Blue Moon

Note that due to the 29-day lunar cycle the exact dates of the full moon move every year. Most seasons have three full moons, but because of the variation some seasons have four full moons. The term "blue moon" was used to identify one of these extra full moons. A mistaken definition in the March 1946 edition of Sky and Telescope magazine claimed the blue moon fell on the second full moon of the calendar month. This mistake caused widespread misunderstanding until it was finally corrected in 1999.

Q158. What is a blue moon?

A158. A blue moon occurs when there are four full moons in any given season, rather than the usual three. The phrase has nothing to do with the actual colour of the moon.

We have four seasons in a year, each about three months long. So, most seasons have three full moons. But because of the lunar cycle at 29.5 days, is shorter than our calendar months, a fourth full moon occasionally appears in a season. So, according to the Farmers' Almanac definition, the blue moon is the third full moon in a season with four full moons. A mistaken definition in the March 1946 edition of Sky and Telescope magazine claimed the blue moon fell on the second full moon of the calendar month. This mistake caused widespread misunderstanding until it was finally corrected in 1999.

One lunation (an average lunar cycle) is 29.53 days. There are about 365.24 days in a year. Therefore, about 12.37 lunations occur in a year. There are 12 months in a year and normally there is one full moon each month. Each calendar year contains roughly 11 days more than the number of days in 12 lunar cycles. The extra days accumulate, so every 2.716 years (2 years, 8 months, and 18 days) there is an extra full moon in the year. The extra full moon necessarily falls in one of the four seasons, giving that season four full moons instead of the usual three, and, hence, a "blue" moon.

Sometimes we have a calendar month without a full moon. This can occur only in February. It happens about every 19 years. When February is without a full moon, then the preceding January or December and the following March or April have two full moons.

Notes:

1. This definition of blue moon is what gave rise to the phrase “once in a blue moon,”.
2. A black-moon is supposedly the flip side of a blue-moon: the third new moon in a season that has four new moons.

Q159. On a full moon, the earth is between the sun and the moon. Then, how does the sunlight get to the moon?

A159. The moon's orbit around earth is tipped about 5 degrees to earth's orbit around the sun. This means that the moon spends most of the time either above or below the plane of earth's orbit, where the earth's shadows lie. That is why earth seldom blocks the sunlight from reflecting off the full moon. Occasionally, though, earth does get directly in the way, and we have a lunar eclipse. (More on this later)

Q160. How is it that during a thin crescent moon one can sometimes see the outline of the rest of the moon?

A160. This is due to “Earthshine”: the sunlight which is reflected from earth to the moon. In the lunar-night, the moon-ground is lit up by the Earthshine. A crescent moon means that the sun, earth, and the moon are lined up, which means that earth is at its full-earth phase when seen from the moon’s surface. The earth, at that moment, appears 13 times larger, 3.4 times more reflective and a total of 43 times brighter than the full moon does from earth. So even when the sun is not out at all, there is still plenty of light on the moon’s surface. Therefore, when the moon is in a crescent phase, one can still see features on the dark portion of the moon.

Crescent Moon. The remaining disk of the moon is still illuminated by the light reflected from earth (earth Shine) that’s then incident upon the lunar surface.
Image Credit: NASA/Bill Dunford
Updated: May 1, 2019
https://solarsystem.nasa.gov/resources/2425/bright-moon-and-crescent-venus/

Q161. How is it that we can sometimes see the moon during the day?

A161. There is a misconception that the moon is only visible at night. The reality is that we frequently see the moon in the day. The only phases of the moon that cannot be seen in the day are:

(1). **Full moon:** when the moon is opposite the sun in the sky (which is when full moon happens), it will rise as the sun sets and set as the sun rises. Therefore, a full moon will be up all night long, but not during the day.

(2). New moon which is not visible from earth at all.

The moon can sometimes appear in the sky during the daytime because it is the closest object to the earth. The moon is approximately as bright in reflected sunlight as the sky during the day, and that's why we can see the moon during the day, but not the stars during the day because the stars are nowhere near as bright as the blue sky during the day. But we don't see the moon all the time during the day because of where the moon might be in the sky.

Q162. What is a Super Moon and a Micro Moon?

A162. Super moon is not an astronomical term and was coined by an American astrologer Richard Nolle in 1979.

The Moon's orbit around the earth is elliptical, which means that sometimes it is closer to earth (known as perigee) and sometimes far (apogee). A super moon is a full moon at the perigee (near the earth). Since it is closer to earth, therefore appears big and bright in the sky. According to NASA, from earth a Supermoon will appear approximately 14 per cent larger and 30 per cent brighter in the sky. But every full moon is not a super moon. From one full-moon to another full-moon it is about one month or, 29.53 days. However, the moon completes one round around the earth in 27.3 days. Due to this mismatch in timing, a super-moon does not take place on every full-moon day. Of the possible 12 or 13 full moons each year, usually three or four may be classified as supermoons.

When a full moon occurs at apogee (far from earth), it's a 'micro-moon'.

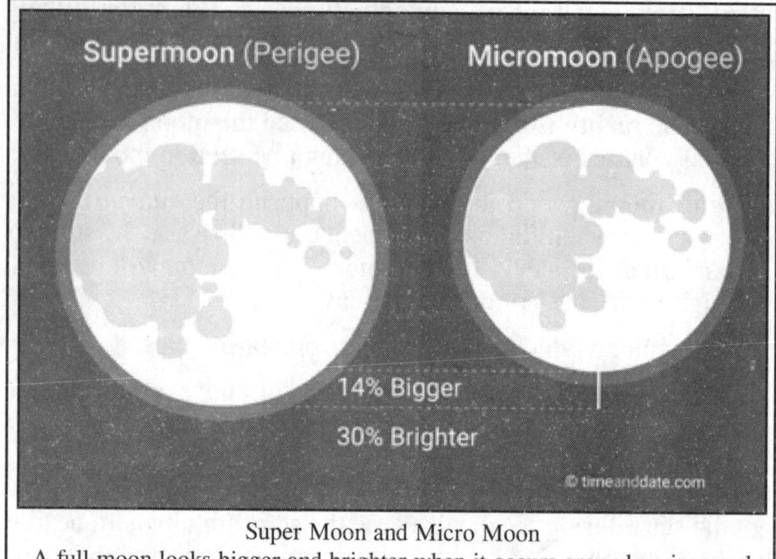

Super Moon and Micro Moon
A full moon looks bigger and brighter when it occurs around perigee and smaller and dimmer at apogee.
Credits: Thorsen, Steffen. timeanddate.com
https://www.tedmontgomery.com/remarks/16.7-12/moon/supermoon.html
Reproduced with permission

Q163. Do the moon phases affect the rain on the earth?

A163. David Need in his blog has mentioned that a recent research, based on 15 years of data collection between 1988 and 2012, has shown that the moon's phases may affect the rain on the earth. When the moon is overhead, its gravity interferes with the earth's atmosphere making it to bulge and the pressure goes up. Scientists say that this effect on the rain is very small, about 1 per cent of the total amount of rainfall. This means that other characteristics of the weather are not changed, and people may not even notice any difference. However, this change can be useful when testing the climate models. Researchers continue their studies at a bigger scale, from normal rain to heavy downpours, to find what rain categories are more prone to be affected by moon's phases. In addition, they want to understand if the moon's phases have any effect on the frequency of rainstorms.

Reference

Need, David. The Moon's Phases Really Do Impact Rainfall on Earth, Study Finds. Science Alert. February 2, 2016. https://www.sciencealert.com/the-moon-really-does-impact-rainfall-on-earth-study-finds

Q164. Does the full moon affect animals' (especially wolves) or people's behaviour?

A164. Wolves are active nocturnally and hunt for food at night and sleep during the day. They howl at night to communicate to other wolves that this is their territory. They also howl to find fellow pack members when they are apart. But researchers have not been able to confirm that wolves howl any more when the moon is full than during any other night.

Regarding people, no one has ever been able to show a conclusive connection between the full-moon and human biology or behaviour. The ancient philosophers of Greece and Rome postulated a watery-brain theory according to which, the water in the brain might create a wave (just like tidal motion in the sea) of strange behaviour whenever the moon was particularly full or large in the sky. Even today, some psychiatrists continue to hold up this theory. But this theory doesn't hold any standing. The gravitational effect of the moon is just as potent during new moons, when the moon is invisible to us, as it is during full moons.

Q165. Why does the moon seem yellow, orange, or red when it is near the horizon?

A165. This phenomenon is due to the scattering of light by earth's atmosphere. While in reflection, light is deflected in one direction, but in scattering it gets deflected in all directions.

Some particles and molecules found in the atmosphere, including particulates made by industry, can scatter light in all directions.

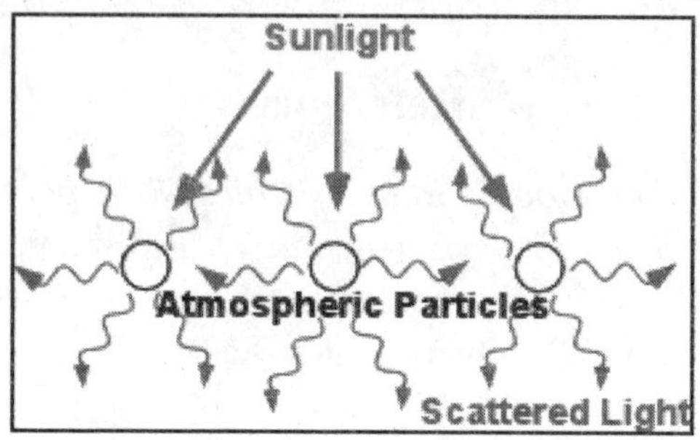

When the moon is near the horizon, moonlight pass through much more atmosphere than when it is directly overhead. By the time moonlight reaches our eyes, the blue, green, and purple colors (smaller wavelengths) of visible light have been scattered away by air molecules. And we see yellow, orange, or red. Sometimes the moon appears orange even when it's directly overhead. This occurs when there's a lot of dust, smoke, or pollution in the atmosphere. The size of those particles will determine the type of colour we will see.

Q166. Just like we see the moon's phases from earth, are earth's phases also visible from the moon's surface?

A166. Yes! Just as the moon shines onto the earth and brightens our night, the earth shines onto the moon's sky. But remember, only one side (near side) of the moon always faces us. So, we would have to be on that side to see the earth.

The earth is over forty times brighter than the moon. The earth's diameter is about 3.67 times the diameter of the moon, which means it appears thirteen and half times bigger than the moon in the sky. Another factor is reflectivity referred to as albedo (the ratio of the light reflected by an astronomical body to that received by it). The moon and earth both get their brightness by reflecting sunlight. The average albedo of the moon is 0.12 while that of earth is 0.33.

The earth and moon, to scale, in terms of both size and albedo/reflectivity. Note how much shiner the earth is compared to the moon. A "full earth" as seen from the moon is about 43 times brighter than the full moon is as seen from earth.
https://bigthink.com/starts-with-a-bang/ask-ethan-how-bright-is-the-earth-as-seen-from-the-moon/
Credits: NASA / Apollo 17.

Just like the moon appears in our night sky from month to month in phases: new, waxing, full, waning, and new again. And so, the earth seen from the moon also appears in phases: full earth, waning earth, new earth, waxing earth and full earth again. The phases are always the reverse of each other. When we see the moon as nearly full, any moon people would see a slim crescent earth. When we see a completely full moon - the moon, earth, and sun are in a line, with earth in the middle. Then people on the moon would not see earth at all because it would be hidden in the sun's glare.

Also, the moon is slow in rotation compared to earth's rotation. The moon appears to rise and set as seen from earth not because of the moon's motion, but because earth spins once a day on its axis. Because one side of the moon always faces us, from most places on the moon, earth does not appear to rise or set. Instead, from a given point on the moon's near side, you would always see earth hanging in moon's sky.

Earth in the Moon's Sky
December 12, 1972, during Apollo 17 mission's first EVA (Extravehicular Activity), Eugene Cernan photographed Harrison Schmitt deploying the American flag with earth in the background in the moon's sky.
https://skyandtelescope.org/astronomy-blogs/explore-night-bob-king/observing-earth-from-the-moon/
Credit: NASA

❑

CHAPTER 14

Celestial Shadows: Transit, Eclipse, and Occultation

Q167. What is an eclipse?

A167. The English word "eclipse" comes from the Greek meaning disappearance, abandonment. An eclipse occurs when one object in space blocks an observer from seeing another object in space. Anytime when there are three bodies (example sun, earth, and the moon) lined up so that one blocks the light from another, we call that an eclipse. The moon is eclipsed when it is within earth's shadow.

Q168. What are umbra and penumbra?

A168. Shadows cast by the moon and the earth are two-part shadows, a dark part and a light part called umbra and penumbra respectively. Umbra is inner core of total darkness; the disc of the sun is completely blocked. Penumbra is outer, partial shadow, sun's disc is only partly blocked, with a bit peeking over the edge.

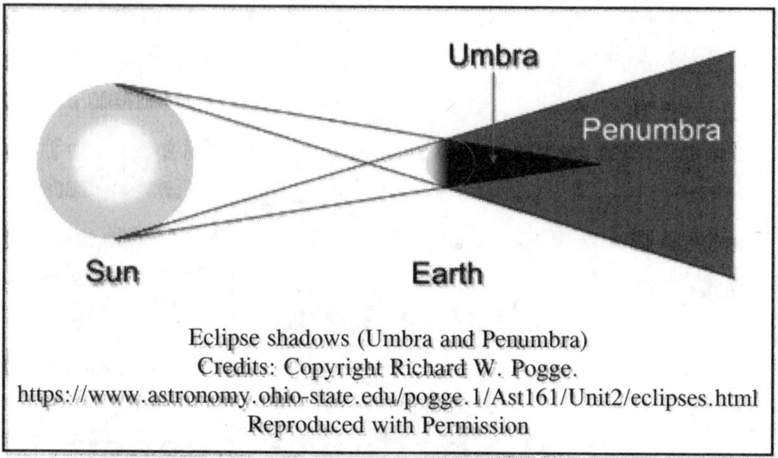

Eclipse shadows (Umbra and Penumbra)
Credits: Copyright Richard W. Pogge.
https://www.astronomy.ohio-state.edu/pogge.1/Ast161/Unit2/eclipses.html
Reproduced with Permission

Q169. What are lunar and solar eclipses?

A169. From earth there are two main types of eclipses: solar eclipses and lunar eclipses. At special times during any year, sun, earth, and the moon "line up".

Lunar Eclipses occur when the moon is full, and earth blocks sun's rays reaching to the moon. The moon passes through earth's shadow. While solar eclipses occur on the "New moon", whereby the moon moves between earth and sun. The earth passes through the moon's shadow.

> **Note:** Other kinds of eclipses happen too. For instance, imagine you are standing on the surface of Jupiter, you might see one of its moons eclipsing the sun.

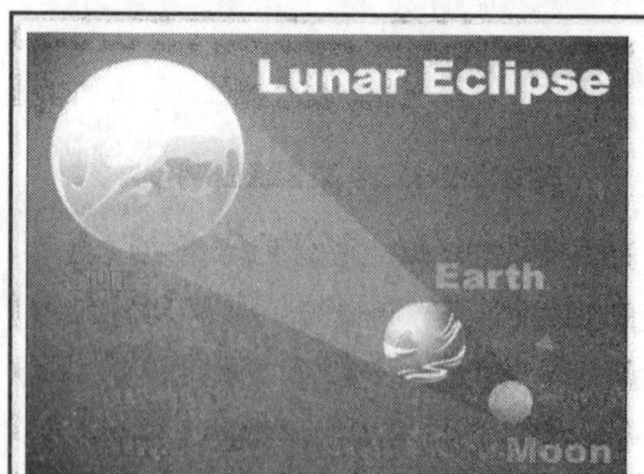

Lunar Eclipse
Credit: NASA
https://www.nasa.gov/audience/forstudents/k-4/stories/lunar-eclipse-diagram
Last Updated: January 11, 2017

Solar Eclipse
Credit: NASA
https://www.nasa.gov/content/eclipses-and-transits-overview
Last Updated: August 7, 2017

Q170. What are the different types of lunar eclipses?

A170. Lunar Eclipse Can Take Different Forms.

A penumbral eclipse is when the moon passes through earth's penumbral shadow. This is difficult to see. The penumbra causes a subtle darkening of the moon's surface. Normal human eyes can't perceive the difference.

A total lunar eclipse occurs when the whole of the moon passes deep through earth's umbral shadow.

When only a part of the moon enters earth's shadow, the event is called a partial lunar eclipse.

The moon's speed is about one kilometre per second. The total time between the moon's first and last contacts with the earth's shadow could last up to four hours and the totality may last up to 107 minutes.

Different forms of lunar eclipses
Credits: Copyright Richard W. Pogge.
https://www.astronomy.ohio-state.edu/pogge.1/Ast161/Unit2/Images/lunar.gif
Reproduced with Permission

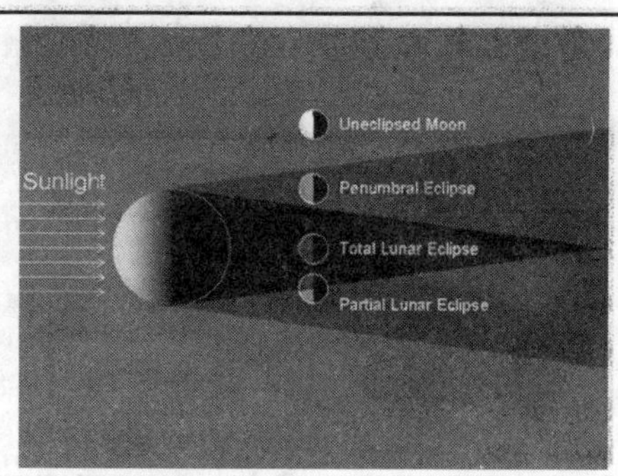

Q171. Why does the moon look red during a total lunar eclipse?

A171. The same phenomenon that makes our sky blue and our sunsets red causes the moon to turn red during a lunar eclipse. It's called Rayleigh scattering. Light travels in waves, and different colours of light have different physical properties. Blue light has a shorter wavelength and is scattered more easily by particles in earth's atmosphere than red light, which has a longer wavelength. Red light travels more directly through the atmosphere.

During a lunar eclipse, the moon is in earth's shadow and should appear black, but some sunlight reaches there after getting refracted from earth's atmosphere. The sunlight first passes through the earth's atmosphere which scatters most of the blue coloured light. The remaining light is a deep red or orange in colour. earth's atmosphere also bends (refracts) some of this light so that a small fraction of it can reach and illuminate the moon. If the earth had no atmosphere, then the moon would be completely black during a total eclipse. The more dust or clouds in earth's atmosphere during the eclipse, the redder the moon will appear.

Total and Partial Lunar Eclipses
Credits: Copyright Richard W. Pogge
https://www.astronomy.ohio-state.edu/pogge.1/TeachRes/Artwork/Sky/Lunar_partfull.gif
Reproduced with permission

Q172. What are the different forms of solar eclipses?

A172. There are four types of solar eclipses: total, partial, annual and hybrid.

Total: During a total solar eclipse, the observer is inside the moon's umbra and the moon completely covers the sun. Total solar eclipses are localized and short and the totality lasts at most about 7.5 minutes with the shadow sweeping rapidly west to east across the earth's surface.

Total solar eclipses provide a rare chance to see the sun's atmosphere, called the corona. The corona is very dim, and it's usually hard to see because the sun is so much brighter. However, during those few minutes of a total solar eclipse, all you can see is the light from the corona.

Partial: During a partial solar eclipse, the observer is inside the moon's penumbra and only part of the sun is obscured.

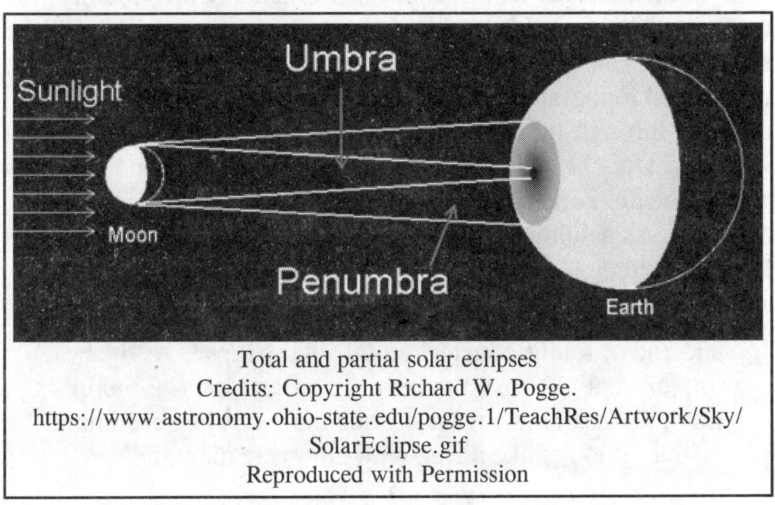

Total and partial solar eclipses
Credits: Copyright Richard W. Pogge.
https://www.astronomy.ohio-state.edu/pogge.1/TeachRes/Artwork/Sky/SolarEclipse.gif
Reproduced with Permission

Annular: An annular eclipse is when the moon doesn't cover the entirety of the sun, even if the moon's is completely centered in front of it. This happens when the moon is farthest from earth (at or near apogee). The moon's umbra does not touch the earth and observers in the shadow path see the sun as a ring ("annulus"). The bright ring around the moon is called the "ring of fire."

Annular Solar Eclipse
May 19, 2012
Image Credit: Photograph by Mikael Svalgaard, www.leif.org/mikael
https://apod.nasa.gov/apod/ap120519.html
Reproduced with permission

Hybrid: A hybrid eclipse is a combination of a total and an annular eclipse. It starts as one type and transitions to another.

Q173. What is the diamond ring effect in a solar eclipse?

A173. The Baily's beads effect or diamond ring effect is a feature of total and annular solar eclipses. Moments before totality, the moon will be covering most of the sun and the corona will be seen like a ring around its edges. As the moon covers the sun, the rugged topography of the lunar limb allows beads of sunlight to shine through in some places while not in others. The effect is named after Francis Baily, who explained the phenomenon in 1836. The diamond ring effect is seen when only one bead is left, appearing as a shining "diamond" set in a bright ring around the lunar silhouette.

Note: The diamond-ring effect occurred at the beginning and end of totality during a total solar eclipse. As the last bits of sunlight pass through the valleys on the moon's limb, and the faint corona around the sun is just becoming visible, it looks like a ring with glittering diamonds on it.

Q174. The sun is much larger than the moon, then how do we get a total solar eclipse?

A174. The sun is approximately 400 times larger than that of the moon. What a coincidence that it also lies roughly 400 times farther away. This means both disks, sun, and the moon, appear to be the same size.

An illustration of a total solar eclipse. The diamond-ring effect Image credit: NASA/GSFC/CI Lab https://spaceplace.nasa.gov/total-solar-eclipse/en/

Q175. Why are some of the solar eclipses of longer duration than others?

A175. The reason is because the earth is not always at the same distance from the sun and the moon is not always at the same distance from earth. The earth-sun distance varies by 3 per cent and the moon-earth distance by 12 per cent. The result is that the moon's apparent diameter can range from 7 per cent larger to 10 per cent smaller than the sun.

Q176. Why don't we have eclipses on each Full Moon and New Moon?

A176. The solar system is rather flat with all its celestial bodies moving in nearly the same plane not too far from the ecliptic plane. The moon's orbit is tilted with respect to the ecliptic by about 5 degrees. Half the time the moon is north of the ecliptic and half the time south. The points where the moon's orbit passes through the ecliptic are called lunar nodes. The ascending or north node is where the moon crosses from south of the ecliptic to north of the ecliptic. The descending or south node is where the moon crosses from north of the ecliptic to south of the ecliptic. About every six months, when the sun and the moon are close enough to a lunar node to align with the earth, there is the eclipse season, which lasts around 34.5 days. An eclipse can only occur during an eclipse season. In every eclipse season, there are 2 to 3 eclipses, and at least 1 of these is always a solar eclipse, at the most 2.

If the shadow of the moon hits the earth, the sun is eclipsed in the shadow area (solar eclipse). If on the other hand, the shadow of the earth covers the moon, we have an eclipse of the moon. Either of these can only happen when the sun, earth and moon are on the same straight line.

It takes close to a month for the moon to go around the earth and during that time its orbit crosses the ecliptic twice, as it goes from one side to the other. At the time of crossing, the sun may be anywhere along the ecliptic; usually it is not on the earth-moon line, and therefore an eclipse usually does not take place. Occasionally, however, it is on that line or close to it, then there is an eclipse. Had the moon's orbit been exactly aligned with the ecliptic, there would have been a solar eclipse at every new moon and a lunar eclipse at every full moon.

Note: The nodes have different names across many cultures around the world. In ancient European texts, the ascending node is referred to as the dragon's head and the descending node as the dragon's tail.

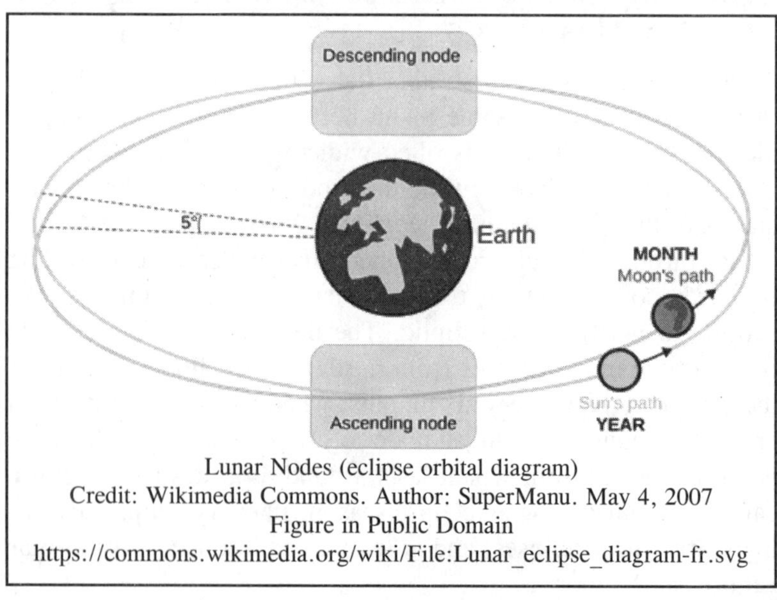

Lunar Nodes (eclipse orbital diagram)
Credit: Wikimedia Commons. Author: SuperManu. May 4, 2007
Figure in Public Domain
https://commons.wikimedia.org/wiki/File:Lunar_eclipse_diagram-fr.svg

Q177. Why is predicting a lunar eclipse easier than a solar eclipse?

A177. The reason is simple, it is how large is the shadow. The diameter of earth's shadow upon the moon is over 12,000 kilometres (7455 miles) during a lunar eclipse. Compare this with the moon's 300 km (185 mi) shadow on the earth during solar eclipses. This makes predicting lunar eclipses easier, even when you do not know the precise details of the moon's orbit. You can be just about anywhere on the side of the earth facing the moon, and still see a lunar eclipse. But solar eclipses are observed from comparatively smaller regions of the earth. So, for a solar eclipse, you must be in a very specific geographic location. To forecast a solar eclipse, you would need to know the details of moon's orbit to a small fraction of a degree of arc to predict where the moon's shadow will cross earth's surface. Else, a predicted solar eclipse may not be visible from your region of the earth.

Q178. What are some amazing facts about eclipses?

A178.

Solar Eclipses

- The longest duration for a total solar eclipse is 7.5 minutes.
- Eclipse shadows travel at 1770 kilometres per hour (1,100 miles per hour) at the equator and up to 8050 kilometres per hour (5,000 miles per hour) near the poles.
- The width of the path of totality is at most 270 kilometres (170 miles wide).
- Each year there are between 2 and 5 solar eclipses.
- Only partial solar eclipses can be observed from the North and South Poles.
- Total solar eclipses are rare, happening only once every 18 months.

- Every eclipse begins at sunrise at some point in its track and ends at sunset about halfway around the world from the start point.
- Partial solar eclipses can be seen up to 4,830 kilometres (3,000 miles) from the track of totality.
- Before the advent of modern atomic clocks, studies of ancient records of solar eclipses allowed astronomers to detect a 0.001 second per century slowing down in earth's rotation.
- Local animals and birds often prepare for sleep or behave confusedly during totality.
- Local temperatures often drop about 10 degrees F (about 5 degrees C) near totality.

Lunar Eclipse

- Some lunar eclipse can last up to 3 hour and 45 minutes.
- Three times a year is the maximum time a lunar eclipse usually occurs.
- During a total lunar eclipse, darkness of the moon varies due to the variation in the composition of earth's atmosphere.

Both – Solar and Lunar Eclipse

- Lunar eclipses are more widely visible because earth casts a much larger shadow on the moon during a lunar eclipse than the moon casts on earth during a solar eclipse. Therefore, more people on earth can view a lunar eclipse than a solar eclipse.
- Usually, eclipses alternate from solar to lunar.
- There are typically more solar eclipses than lunar eclipses per year.
- Nearly identical eclipses (total, annual, or partial) occur after 18 years and 11 days, or every 6,585.32 days (Saros Cycle).

Reference

NASA. Eclipse 99 - Amazing Facts - Little Known Facts About Solar Eclipses. https://science.ksc.nasa.gov/mirrors/gsfc/omni/eclipse99/pages/amazing.html

Q179. What is the Saros cycle and how does it foretell eclipses?

A179. Eclipses were considered scary events by many early civilizations. They wanted to know some methods to predict them. That is why, many centuries ago, Babylonian astronomer-priests studied patterns of eclipses. One of the most brilliant of their observations is the "Saros Cycle". The saros is a period of exactly 223 synodic months, approximately 6585.3211 days, or 18 years 11-1/3 days (10-1/3 days when five leap years are included). After which the sun, earth, and the moon return to nearly the same relative geometry and a nearly identical eclipse will occur. The name "saros" was applied to the eclipse cycle by Edmond Halley in 1691. With the help of Saros Cycles, it became easier to predict eclipses.

In modern times, astronomers have better understandings of the moon's shape, topography of the earth, and relative positions of sun, earth, and the moon. They, therefore, can make more accurate predictions of eclipses including an incredible detailed and accurate accounting of where the eclipse shadow will pass across the world.

Q180. What are transit, eclipse, and occultation?

A180. Transit, eclipse, and occultation are astronomical phenomena. These are all interrelated and occur when one astronomical object casts its shadow on another. These are the events that involve a body, which moves between two others, obscuring some or all the light of one of the bodies as seen from the other.

Transit of Venus across the face of the sun. Sequence of images of 171 transit composited together to show the path of Venus.
Credits: NASA/Goddard Space Flight Center/SDO
https://svs.gsfc.nasa.gov/10996
Date: June 5, 2012

The study of celestial shadows is of great significance in astronomy.

A **transit** occurs when a relatively small body crosses in front of a larger body. For example, Mercury and Venus transiting the sun in silhouette and tracing a straight path across its face; or a satellite transiting its planet.

Note: On June 5, 2012, Solar Dynamics Observatory (SDO) collected images of the transit of Venus across the face of the sun. This event lasted approximately six hours

Occultation refers to the complete or partial obscuring of an astronomical object by another. **Eclipses** are a class of occultations wherein the bodies involved appear to be of similar sizes. When earth is casting shadow on the moon, it is lunar eclipse. And when the moon casts shadow on the earth, it is a solar eclipse. So, one can surmise that during a solar eclipse, the sun is "eclipsed" or, "occulted." The term occultation is most often used to describe the disappearance of a star, nebula, or planet behind the moon. These may be considered a small-scale eclipse event, not as spectacular as a solar or lunar eclipse, but an event which occurs much more frequently. Occultations don't only take place on earth, they can take place anywhere in the universe. For example, the Apollo 12 astronauts flew through earth's shadow and saw a solar eclipse on their way home in 1969.

Q181. What is the significance of studying celestial shadows?

A181. Much of what is known about the universe has come from the study of celestial shadows.

The curved shadow of the earth on the moon's surface during a lunar eclipse revealed that the earth is round.

Shadows were used by the Greek mathematician Eratosthenes to estimate the circumference of the earth. Galileo estimated the heights of the moon's mountains. Some other important discoveries made by observers of these events are relative dimensions of the planet and its host star, orientation of the planets' orbit, temperature and composition of the planetary atmosphere, discovery of new planets, unsuspected rings of planets, discovery and makeup of extra solar planets, evidence of satellites and asteroids, dimensions of asteroids.

The study of solar eclipses has contributed a great deal of knowledge about the chromosphere and corona, the thin external layers of the sun. Study of a solar eclipse of 1868 resulted in the discovery of an unknown gas, Helium, which wasn't identified on earth till 1895. Because of this, it was named after the Greek word for the sun - Helios. The total solar eclipse of May 29, 1919 was used by British astronomer and mathematician, Sir Arthur Eddington to prove Albert Einstein's theory of general relativity. By taking pictures of stars near the sun during totality, Eddington was able to show that gravity can bend light.

Observing the moon's occultations of stars has resulted in the discovery of binary star systems (double stars). In these observations, stars appear to fade in two stages, rather than instantly, thus indicating that there are actually two stars, which may not be clearly evident by telescopic viewing.

A ring system around Uranus was discovered in 1977 while astronomers were observing an occultation of a star by Uranus. In 1984, while Neptune was occulting a star, scientists were able to observe what appeared to be an incomplete ring system.

More recently scientists are using studies of celestial shadows to understand global warming.

And the list goes on.

Q182. What are some of the Myths, Legends, and Folktales About Eclipses?

A182. Several cultures around the world have many folktales, stories, and beliefs about the eclipses. These may include a monster devouring the sun and / or a punishment from the gods for human errors. In ancient China, the earliest word for eclipse, shih, meant to eat, and eclipses were believed to be caused by a dragon eating the sun. In Vietnam, the sun eater was a frog. For the Native American Pomo, it was a bear. In Yugoslavia it was a werewolf, and in Siberia a vampire.

Some of the stories and rituals are given below.

Vedic (Hindu) story Samudramanthan (Story of Eclipses and Rahu-Ketu)

(Slightly different versions are available in literature. Common theme is the same).

Long ago there was a great war between the Asuras (demons, who are another group of super humans who fight with gods for supremacy) and the Devatas (gods). The Devatas were getting defeated. They went to Lord Brahma to seek his help. Brahma told them to churn the ocean to obtain nectar (Amrit), consuming of which would make them immortal. Then they would be able to defeat the Asuras. Churning the ocean was not an easy job so, Devatas sought the help of Asuras. Using the great snake Vasuki as the rope and the mountain Mandrachal as the shaft, they began churning the ocean. Lord Vishnu, in the form of a tortoise, became the base. During the churning, the ocean kept throwing out a variety of items including a pot of Vish (poison) which created havoc in all the three worlds. Lord Shiva drank the poison to save the world, but he did not let the poison go beyond his neck which became blue in colour. That is why Lord Shiva is also called 'NeelKant' ("one with a blue neck"). At last, Dhanvantari came out with a pot of Amrit and immediately there was a mad scramble for it among Devatas and Asuras. Seeing this, Lord Vishnu took the form of Mohini (a beautiful dancer) and offered her services in distributing the Amrit equally to both the Asuras and Devatas. She made both sit in two different rows, but she gave the Amrit to the Devatas only. Towards the end, one of the Asuras, named Swarbhanu, saw through her trick. He went and sat among the Devatas and drank the Amrit. The moment Chandra (Moon) and Surya (Sun), who were sitting beside him, saw that he was an Asura, they informed Mohini. Lord Vishnu came in his real form and let out the Sudarshan Chakra (wheel) at Swarbhanu separating his neck from the body. But Swarbhanu did not die as he had drunk the Amrit. His head was called 'Rahu' and his torso 'Ketu'. Ever since, Rahu and Ketu are very angry with the sun and the moon and now spend all their time trying to take revenge on them. Hence the eclipses occur when Rahu gulps down sun or the moon. However, because Rahu is only a head, the sun, and the moon escape Rahu as he gulps them down.

Since Rahu and Ketu also move in the sky like other planets, they are called dark or unseen planets. Their motion is also predictable and hence the ancient Indian practice is to worship 9 planets - Sun, Moon, Mercury, Venus, Mars, Jupiter, Saturn, Rahu, and Ketu.

Later, when Aryabhata gave the complete formulation of the eclipses, he used the concept of Rahu and Ketu to describe the ascending and descending nodes of the point of intersection of the Lunar and Solar orbits. These are then transformed into specific moving points of a Mythology.

Note: In some texts, the name of the mountain is mentioned as Mainak.

Special Prayer in Islam During Eclipses

When one of the sons of Prophet Muhammad died during the lifetime of the Prophet, there was a solar eclipse on that day. Some people suggested that the sun eclipsed because of the young child's death and the Prophet's sadness on that day. However, the Holy Prophet corrected this for the people and said:

"The sun and the Moon are two signs among the signs of Allah. They do not eclipse because of someone's death or life. So, when you see them, invoke Allah and pray till the eclipse is clear."

As such, during the time of an eclipse (kn̄own as kusuf), Muslims perform a specific prayer called Salatul Ayaat. It is prescribed that during times of fear, people should turn towards Allah and pray to Him and glorify his majesty.

Also, the Occultation in some sector of Shia Islam refers to a belief that the messianic figure, or Mahdi, who in Shia's thought is a male descendant of the founder of Islam, Muhammad, was born but disappeared, and will one day return and fill the world with justice and peace.

Eastern Bolivia

Some tribal communities of eastern Bolivia believe that the sun was a beautiful man, and the moon was his sister. Eclipses were caused by celestial serpents which attacked these luminaries, threatening mankind with darkness. This catastrophe was to be followed by the transformation of men into hairy animals and by their mutual extermination.

Eastern Paraguay and Northern Brazil

They believe that eclipses are caused by the Eternal Bat, or in some cases the Celestial Jaguar, which bites the sun or the moon. People have a very pessimistic outlook on the future of the world. They are firmly convinced that its end is near. Very soon our great father will set the earth on fire, unleashing the Eternal Bat and the Blue Jaguar which will destroy the stars and mankind.

Armenian folklore

As among many other peoples, the eclipse of the sun and moon was thought to be caused by dragons which endeavor to swallow these luminaries. When the moon was at an eclipse, the sorcerers said that it resembled a demon. It was, moreover, a popular belief that a sorcerer could bind the sun and moon in their course or deprive them of their light. Needless to add that the eclipses and the appearance of comets give alert about evil. Their chronologies are full of notices of such astronomical phenomena that presaged great national and universal disasters.

The Legends of the Jews

These legends explicitly link solar eclipses to the expulsion of Adam and Eve from the Garden of Eden and to the crucifixion of Jesus. The sun is eclipsed because it cannot stand tragic happenings in world history.

The ancient Greeks thought of a solar eclipse as an act of abandonment, a terrible crisis, and an existential threat. It meant that the king would fall, that terrible misfortunes would rain down on the world, or that demons had swallowed the sun.

In ancient Egypt, Apep (also spelled Apepi or Aapep) or Apophis, the serpent of chaos and death, opposed Ra, the sun god, and was always trying to reach Ra's sky boat to devour the sun disc — but in the end, Ra was always able to fight him off, and the sun would come back.

❑

CHAPTER 15

Tides

Q183. What are tides?

A183. The word "tide" refers to the deformation of land and water of the earth due to gravitational forces of the moon, and to a lesser extent, of the sun. This deformation results in two "tidal bulges" one on the side of the earth nearest to the moon, and the other on the opposite side.

Every particle on the earth is pulled on by moon's gravity. Because liquids move more easily, the pull of gravity is much more noticeable in liquids than in solids. However, the rock does bend, by as much as 30 centimeters (1 foot). The typical tidal range in the open ocean is about 0.6 metres (2 feet). Closer to the coast, this range is much greater which varies globally and can differ anywhere from near zero to over 11 metres (36 feet). The exact range depends on the volume of water adjacent to the coast and the ocean-bed topography. The height of the tides can also vary during a month because the moon is not always at the same distance from the earth. As the moon's orbit brings it in closer proximity to the earth (perigee), the moon's gravitational forces increase which lead to higher tides. Likewise, when the moon is farther away from the earth (apogee), the tides are not as spectacular. During "Super Moon" periods, very high tides can be seen. The world's largest tidal range of about 16 metres (52.5 feet) occurs in Bay of Fundy, Canada. The United Kingdom regularly experiences tidal ranges up to 15 metres (49 feet) between England and Wales.

Q184. Besides the moon and sun, do other planets influence earth's tides?

A184. No. The tidal force generated by other planets is negligible.

Earth's tides are dominated by the combined effect of the sun and the moon's gravitational pull. But the other planets, since they have a gravitational pull of their own, also have a small effect on the tides, but very very small. It is so small that for all intents and purposes, the effect of the planets on earth's tides is imperceptible (see the table below). Lunar tidal forces are dominant because the moon is so nearby. Tidal stretching is proportional to the inverse cube of the distance to the source of gravity. Thus, nearby objects -- even small ones like the moon -- raise high tides, while distant giants like Jupiter don't produce much of an effect.

While the force of gravity falls off as the square of the distance, the tidal forces decline as its cube. The reason is simply that tidal stretching comes from a difference in the gravitational pull felt on two sides of a body.

After the sun and the moon, Venus is the object in our solar system that produces the biggest tides on earth. This is simply because the earth comes closer to Venus than any other planet. Venus increases the size of our ocean tides by less than 0.005 cm.

So, even if all the planets aligned such that their effects combined the additional force would be insignificant.

Maximum tidal forces of the sun, moon, and planets on the earth

Solar System Object	Relative Tidal Force (assuming 1 for Sun)
Moon	2.1
Sun	1.00
Venus	0.000113
Jupiter	0.0000131
Mars	0.0000023
Mercury	0.0000007
Saturn	0.0000005
Uranus	0.000000001
Neptune	0.000000002
Pluto	0.0000000000001

Note: In this table, adapted from the Griffith Observatory's "Planetary Alignments in 2000", the sun exerts 1 unit of tidal force on the earth; the moon has slightly more than twice the effect of the sun; the other nine planets together with all their moons add only another one five-thousandth as much.

Reference

Interplanetary Low Tide. NASA Science. May 3, 2000 https://science.nasa.gov/science-news/science-at-nasa/2000/ast04may_1m

Q185. Where are the highest tides in the world?

A185. The highest tides are spread throughout the world. They occur because of local geography creating a funnel to concentrate the tide from a large oceanic body of water. Below is a listing of 10 locations of countries from around the world where the largest range of tides (the difference between low and high tide), that have been observed.

S. No.	Station	Mean Range (feet)
1.	Bay of Fundy, Nova Scotia	38.4
2.	Ungava Bay, Quebec	32.0
3.	Port of Bristol, United Kingdom	31.5
4.	Cook Inlet, Alaska	30.3
5.	Rio Gallegos, Argentina	29.0
6.	Hudson Bay, Canada	28.5
7.	Granville, France	28.2
8.	Magellan Strait, Chile	28.0
9.	Jersey, Channel Islands	25.1
10.	Kamchatka, Russia	24.1

Source: NOAA

Reference

Tides and Currents. National Oceanic and Atmospheric Administration (NOAA). FAQ https://www.tidesandcurrents.noaa.gov/faq.html#08

Q186. Why there are high tides and low tides?

A186. The moon's gravitational pull is strongest on the part of the earth directly facing the moon. When that part of the earth happens to include an ocean, the water bulges towards the moon. Water on the opposite side of the earth facing away from the moon also bulges outward. These bulges are known as high tides. When high tides occur, water is drawn away from the area between the high tides, causing low tides. Coastal areas experience two high and two low tides every lunar day. Unlike a solar day, a lunar day is 24 hours and 50 minutes, 50 minutes longer than a solar day. The reason: the moon orbits around the earth in the same direction that the earth rotates around its axis. So, it takes the earth an extra 50 minutes to "catch up" with the moon. Therefore, the time between successive moonrises is 24h

50m. High tides occur 12 hours and 25 minutes apart. It takes six hours and 12.5 minutes for the water at the shore to go from high to low, or from low to high.

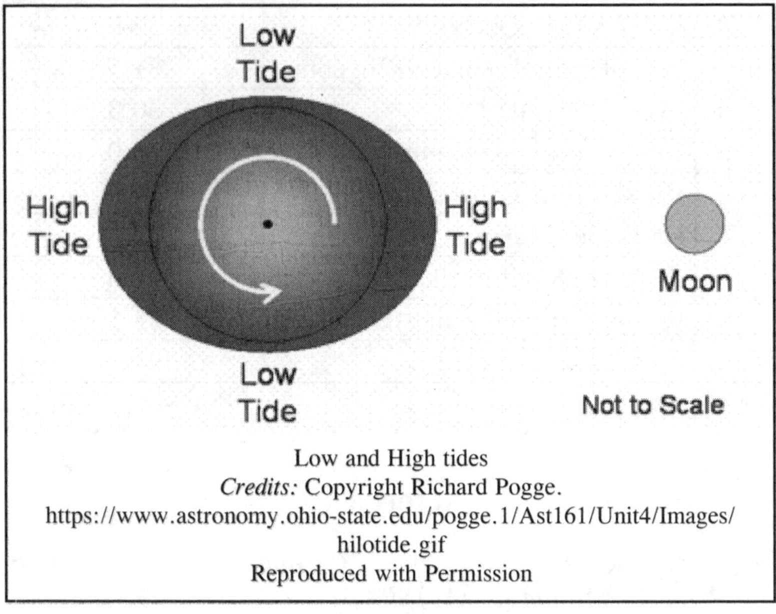

Low and High tides
Credits: Copyright Richard Pogge.
https://www.astronomy.ohio-state.edu/pogge.1/Ast161/Unit4/Images/hilotide.gif
Reproduced with Permission

Q187. What are spring tides and neap tides?

A187. When the sun and the moon are approximately aligned (as they are when the moon is new or full), joint gravitational pulls of the moon and sun add up and we experience especially high tides called "Spring Tides." When there is a 90-degree angle between the two, we get lower tides called "Neap Tides."

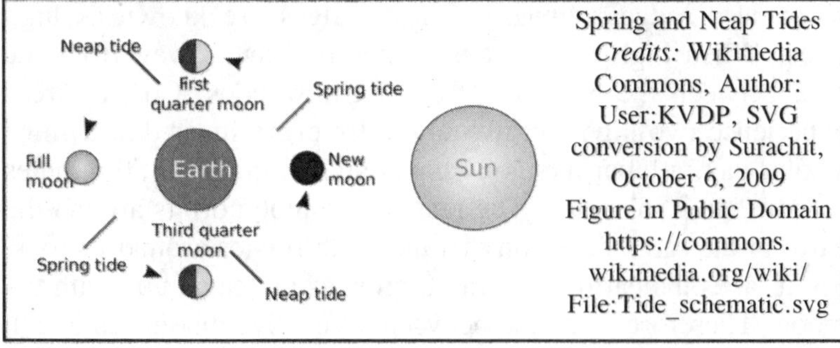

Spring and Neap Tides
Credits: Wikimedia Commons, Author: User:KVDP, SVG conversion by Surachit, October 6, 2009
Figure in Public Domain
https://commons.wikimedia.org/wiki/File:Tide_schematic.svg

Q188. Why there are two high tides each day?

A188. This is a frequently asked question. According to Newton's equilibrium theory, tides result by the difference in the force of gravity between two points. This differential is known as Tidal Force, which is different than the gravitational force. According to Newton's explanation given in 1687, the moon's gravity pulls on the earth and the water on it. The gravitational force on the nearest and farthest sides of the earth is different. The moon is 12,740 km (7,960 miles) closer to the nearest side of the earth than the farthest. This results in a 7 per cent stronger gravitational force on the nearest side compared to the farthest side. This causes a net front-to-back differential gravitational force felt by the earth stretching the earth along the "moon-earth line" while squeezing the earth at right angles to this line. The net result is two tidal bulges on opposite sides of the earth.

Here is how it works:

As mentioned earlier, on the side of earth that is directly facing the moon, the moon's gravitational pull is the strongest. The water on that side is pulled strongly in the direction of the moon. On the side of earth farthest from the moon, the moon's gravitational pull is at its weakest. At the centre of earth is approximately the average of the moon's gravitational pull on the whole planet.

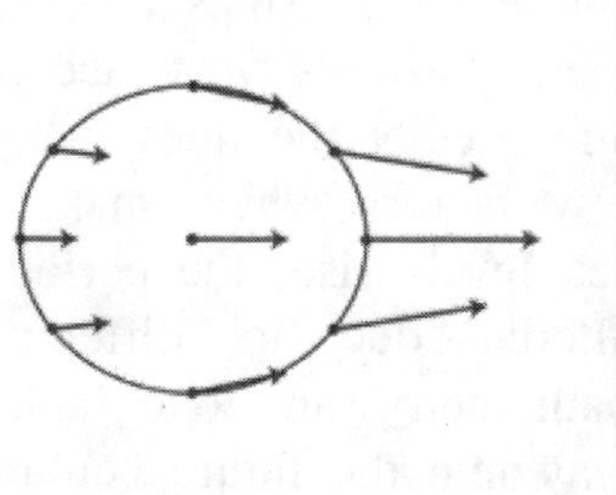

Arrows represent the force of the moon's gravitational pull on the earth. To get the tidal force—the force that causes the tides—we subtract this average gravitational pull on the earth from the gravitational pull at each location on earth.
Credits: NOAA/JPL-Caltech
https://scijinks.gov/tides/
Reproduced with permission

To get the tidal force—the force that causes the tides—we subtract this average gravitational pull on the earth from the gravitational pull at each location on earth. The result is stretching and squashing of earth. This is what causes the two tidal bulges.

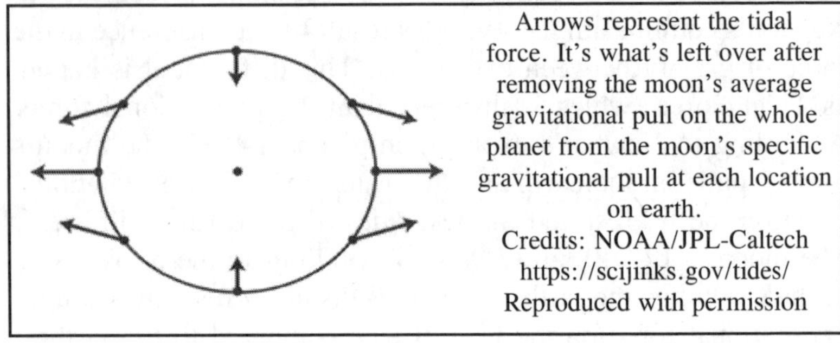

Arrows represent the tidal force. It's what's left over after removing the moon's average gravitational pull on the whole planet from the moon's specific gravitational pull at each location on earth.
Credits: NOAA/JPL-Caltech
https://scijinks.gov/tides/
Reproduced with permission

These two bulges explain why in one day there are two high tides and two low tides, as the earth's surface rotates through each of the bulges once a day.

Reference

NOAA SciJinks. What causes tides?
https://scijinks.gov/tides/

Q189. How far ahead can tides be predicted?

A189. Tides can be predicted far in advance. However, accuracy depends on many factors. Tides are formed by the orbital relationships between earth, sun, and the moon. These relationships are very well understood. We also know their positions very accurately. Therefore, we can compute and predict the tides many years ahead. However, there are few factors which make predictability slightly difficult. As sea levels rise, the periodicity and range of the tide will be altered due to different underwater depth (known as bathymetry) and topography. Therefore, predicting tides a long way into the future could be less accurate. Storm surges are short-term sea-level changes caused by the weather that also affect tidal predictability. Storm surges can only be forecast with the same time horizon as weather forecasting, about two to five days.

Q190. What is tidal braking.

A190. Tidal braking is the phenomenon that causes a body to slow down because of the interaction with an object orbiting it.

Let us understand this between the moon and earth. The moon create tidal braking to the earth causing the earth's rotation to slow down. As we know, the earth takes 24 hours to make one round around its axis and the moon takes about 27 days in making one orbit around the earth. This means that earth's rotational speed is faster than the moon's orbital speed. This results in ocean tides leading the moon by about 10-degrees and creating a friction between the ocean and the seabed. This friction causes the earth to reduce its rotational rate slowly and persistently and thereby lengthening the earth day by 0.002 seconds every century. This does not sound very much but over a period this adds up. This friction energy doesn't disappear and gets transferred, in accordance with the law of conservation of energy, by imparting momentum to the moon. That creates lunar recession whereby, as the earth's rotation slows down, the moon accelerate in its orbit, moving farther from the earth at a rate of about 4 cm per year.

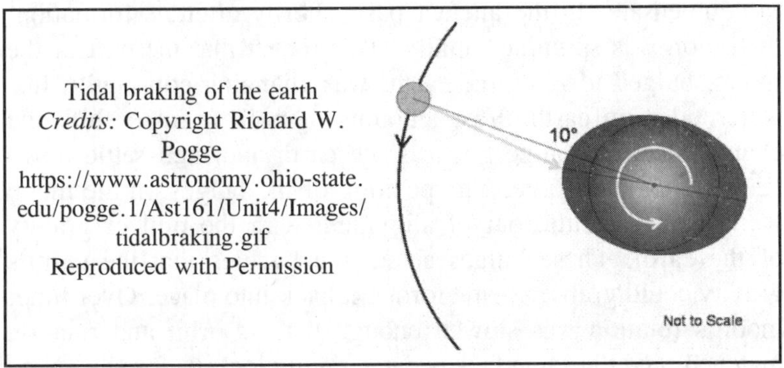

Tidal braking of the earth
Credits: Copyright Richard W. Pogge
https://www.astronomy.ohio-state.edu/pogge.1/Ast161/Unit4/Images/tidalbraking.gif
Reproduced with Permission

Now, imagine the future scenario. Tidal braking will cease eventually when the moon's orbit (earth's month) and the earth's rotation (earth's day) both equal 47 current days in length. The moon will be about 50 per cent farther away from the earth. Once the moon is that far away, it will lose its ability to stabilize the earth on its axis which would begin to tumble in its orbit and the climate would never be stable again. But nothing to worry about, it will be only after about 3 billion years from now.

Q191. What is tidal locking?

A191. Tidal locking is the situation when an object's orbital period matches its rotational period. It is also called synchronous rotation. It is due to the gravitational interaction between objects. Under tidal locking, one side of an astronomical body always faces another.

The moon is tidally locked with the earth and therefore the moon rotates once on its axis exactly as long as it takes to orbit once around the earth, about 27 days. This results in the moon's same face always facing the earth.

Just like the moon's gravity affects ocean tides on the earth, gravity of the earth affects the moon. The earth is also pulling the moon. The pull is even more powerful because the earth is more massive. The gravity from the earth distorts the moon into an oblong shape. The sides pointed towards and away from the earth bulge outward, while the others are pulled inward to compensate. In the ancient past, shortly after its formation, the moon was spinning rapidly. This meant that the part of the moon, bulged towards the earth, was changing constantly, like water tides on earth. Vast amounts of rock need to shift and change shape to bulge towards the earth and then settle down again. This takes time. The position of the bulges on the moon were always a little out of alignment with the pull of gravity of the earth. These bulges acted like handles that the earth's gravity could grab onto and torque it back into place. Over time, moon's rotation was slowed enough that its orbit and rotation matched, and the same face became tidally locked, forever. Now the process is continuing to make the earth tidally locked to the moon as well. In about 40 /50 billion years, the earth and the moon will finally be tidally locked to each other.

The moon is not the only one in the solar system where this happens. This same process happened on all the large moons in the solar system. All major moons of Jupiter and Saturn show the same face to their parent planets. In the case of Pluto and its moon, Charon, both are locked to each other, facing one another for all eternity.

Q192. How can we use tidal energy?

A192. Tidal phenomena are extremely important. Tidal energy is a renewable resource like wind and sun. Energy obtained from tides can be converted into useful forms of power and has potential for future electricity generation for homes and industries. One of the advantages of tides is that they are more predictable than the wind and the sun.

Globally there are only four countries which are generating power through tidal energy. The world's first large-scale tidal power plant was in France (Rance Tidal Power Station, peak output 240 MW) which became operational in 1966. It was the largest tidal power station in terms of output until South Korean plant (Sihwa Lake Tidal Power Station) opened in August 2011 generating 254 MW. The other places where tidal power is used to generate electricity include Canada, China, and Russia. No technology has yet emerged as the clear standard and a large variety of designs are being experimented with.

India has a long coastline with the estuaries (bodies of water, usually found where rivers meet the sea) and gulfs where tides are strong enough to move turbines for electrical power generation. But there are no tidal energy-based power plants in India yet. It is estimated that India has a potential to develop more than 8000 MW of electricity via tidal power (in Gujrat and Sundarbans). In 2011, there was an announcement that a tidal power plant was under consideration in the Gulf of Kutch in Gujarat.

❑

PART III: EXPLORATION

"Curiosity is the essence of our existence."

– Gene Cernan.

CHAPTER 16

Science of Space Travel

Q193. Which science laws are used in space travel?

A193. Newton's law of universal gravitation, Newton's three laws of motion, Kepler's three laws of orbital motion, and Tsiolkovsky rocket equation are used in space travel.

Newton's Laws govern 99 per cent or more of our everyday experiences. They are excellent approximations at the scales and speeds of everyday life, including space travel. However, they are inappropriate for use in certain circumstances, most notably at very small scales (atomic and subatomic particles), very high speeds (approaching speed of light) or very strong gravitational fields (like black holes). Explanation of these phenomena requires more sophisticated physical theories, including general relativity and quantum theory. These are outside the scope of this book.

Q194. What are Newton's law of universal gravitation and three laws of motion?

A194. The fall of an apple inspired Newton to think about how gravity works. He thought that if gravity can reach all the way to an apple tree, could it go higher, as high as to the moon? And if so, it could prove that moon's motion around earth is due to gravitational forces. Newton also showed that the motion of the objects on the earth and of heavenly bodies is governed by the same set of natural laws. According to him an apple falling to earth and the moon falling in its orbit around the earth follow the same rules. Newton realized that the moon constantly falls towards the earth, but due to its fast forward motion it keeps falling past earth. That's how the moon maintains a constant orbit instead of colliding with the earth. On the other hand, in the absence of gravitational attraction the moon would fly off in a straight line rather than orbiting the earth.

Newton's Laws are considered by many to be the most important laws of all physical sciences.

Before we discuss these laws, let us recognize that Einstein's theory of general relativity is more accurate description of gravity. So why do we not use Einstein's theory rather than Newton's. It is not that science does not universally accept general relativity. It is that in most practical scenarios, the corrections due to relativity are so tiny, it is entirely sufficient to use the much simpler Newtonian theory (simpler in part because the mathematics involved is much more elementary) for practical calculations. Einstein didn't invalidate Newton. His theory refined Newton's theory and did so in a manner that Newton himself (who was greatly troubled by the "action-at-a-distance" nature of his theory) would have appreciated.

Newton's Law of Universal Gravitation

Newton defined gravity as a force that acts at a distance and attracts bodies of matter toward each other. This is a universal law, meaning, it describes the behaviour of the material world both in the sky and on earth. Any two objects in the universe attract each other without any exception. The strength of the force depends on the masses of the objects involved. That is why we feel the gravity force of earth and not by other objects like the book we are holding in our hands. The gravitational force also depends on the distance between the objects. The farther apart the objects are, the weaker the force.

According to the Newton's Law of Universal Gravitation, the attraction force between two objects is proportional to their masses and inversely proportional to the square of distance between them. It is also known as Inverse Square Law. In other words, if an object is moved out twice as far from another object, it will feel one quarter of the force; if it is put three times as far away, it will feel one-ninth of the force and so on.

Newton's Laws of Motion

These laws explain how the concepts of force and motion work. Why does an object move like it does, how does the object accelerate or decelerate, how the moon orbits the earth, and the planets orbit the sun, how a person rides a bicycle, or a rocket launches a satellite into space?

Ancient Greeks, including Aristotle, thought that a force must act on an object to remain in motion, the greater the force on a body the greater its speed. Galileo questioned the idea. He developed the concept of inertia, a property of all matter to remain in constant motion or at constant rest unless a force is applied. Galileo reasoned that moving objects eventually stop because of a force called friction. This told astronomers that everything is in constant motion in universe as there is no friction force in space. Galileo suggested that once an object is in motion, no force (pull or push) is required to keep it moving. Force is required only to change its motion.

Newton built upon Galileo's ideas. He discovered three laws of motion concerning the behaviour of moving objects. He explained the way in which objects respond to forces other than gravity. Newton's three laws explain how and why objects move when forces are applied to them.

Newton's First Law of Motion Also Known as Law of Inertia

Newton's First Law of Motion explains the connection between force and motion. According to this law, every object in a state of uniform motion (including zero motion, i.e. state of rest) tends to remain in that state unless an external force is applied to it. This means:

(a) If an object is at rest, it will stay at rest unless acted upon by some force.

(b) If an object is moving at a constant velocity, it will continue at that velocity unless acted upon by some force along the line of motion, and

(c) If an object is moving, it will move in a straight line unless acted upon at an angle by some force.

Newton's Second Law of Motion also known as Law of Dynamics

The first law identified forces as the actions required to stop or start motion, it didn't quantify those forces. Newton's second law supplied the missing link by relating force to acceleration (change in velocity with respect to time). According to this law when a force acts on an object, the object accelerates in the direction of the force. If the mass of an object is held constant, increasing force will increase acceleration. If the force on an object remains constant, increasing mass will decrease acceleration. In other words, force and acceleration are directly proportional, while mass and acceleration are inversely proportional. Another way to express this law is that the net force on the object (say F) is equal to the product of objects own mass (m) and acceleration the object experiences (a).

Force = mass x acceleration

As long as the force continues to operate on a moving object, the object will accelerate, changing its direction or its speed or both. When we take away the force, the object continues to move at a steady speed in a straight line at the new velocity that has built up during the time it was accelerating.

Gravity is a unique type of force that imparts the same acceleration to all objects it affects.

It is known as acceleration due to gravity, normally written as "g". Near earth's surface its value is about 9.8 metres/sec/sec (32 feet/sec/sec). This means that for every second an object falls is falling, its velocity increases by 9.8 metres per second (m/s). Suppose we drop an object from the top of a building. Its initial velocity is zero m/s. At the end of first second of falling, its velocity will be 9.8 m/s; after 2 seconds it will be 19.6 m/s (9.8 m/s + 9.8 m/s) and so on. However, the value of g starts to vary at high altitudes.

When the only force acting on a falling object is gravity, the object is said to be in "free fall".

Newton's Third Law of Motion also known as Action Reaction Law

Newton realized that forces are not "one sided". Whenever an object exerts a force on a second object, the second object also exerts a force back on the first object. The force exerted by the second object is equal in strength and opposite in direction to the first force. The first force is called "action" while the second "reaction'. Newton's third law of motion can be stated as: For every action there is an equal and opposite reaction.

A rocket works on the same principle – it throws material out from its exhaust in one direction, and the reaction pushes the rocket along in the opposite direction.

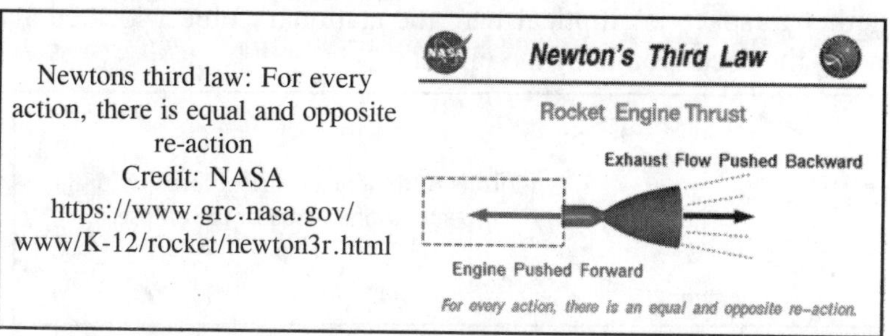

Newtons third law: For every action, there is equal and opposite re-action
Credit: NASA
https://www.grc.nasa.gov/www/K-12/rocket/newton3r.html

Q195. What are Kepler's laws of orbital motion?

A195. Tycho Brahe and Johannes Kepler's contribution to our understanding of planetary motion are very valuable for space travel. Tycho's observations were accurate enough for Kepler to discover his famous three laws of planetary motion. For more than thirty years – mostly in Denmark and then in Prague Tycho had meticulously observed the movements of the planets and the positions of the stars. After Tycho's sudden death, Johannes Kepler got his data, and worked with it for nine more years and found his three laws of planetary motion. These laws are still considered an accurate description of the motion of any planet and any satellite. These are the laws that Newton was able to use to establish universal gravitation.

But it was not easy for Kepler.

There was no hi-tech computation techniques available in those days. Kepler had to work very hard for ten years to empirically fit the data and to come up with his famous laws.

Kepler found that motions of all heavenly bodies follow the same mathematical rules. The planets travel around the sun in elliptical orbits and not circular, as thought by Copernicus. Each planet moves more slowly when it is farther from the sun, then speeds up as it moves closer. Kepler formulated the following three laws explaining orbital motion in space.

Kepler's First Law of Orbital Motion also known as the Law of Ellipses

According to this law, the orbit of an object moving around another in space is elliptical with the stationary object located at one of the focal points of the ellipse.

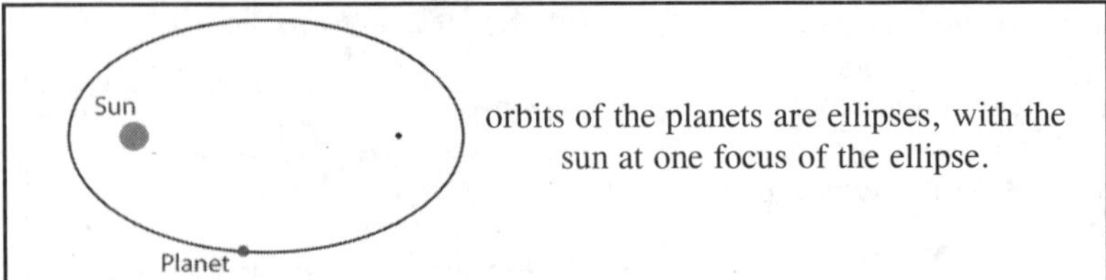

orbits of the planets are ellipses, with the sun at one focus of the ellipse.

Let us consider our sun and earth. earth is orbiting around the sun in an orbital path. The sun is not at the centre of the ellipse, but it is at one focus (generally there is nothing at the other focus of the ellipse). Due to elliptical path the earth-sun distance is constantly changing as the earth goes around.

Kepler's Second Law of Orbital Motion also known as the law of equal areas

This law states that the orbiting object will speed up when it gets closer to the object what it is orbiting about. This is due to the increased effect of gravity.

Alternately, this law for the moon and earth can also be described as: if an imaginary line is drawn from the centre of a satellite (say moon) to the centre of the planet (say earth) that line would sweep out the same area in equal periods of time. This is shown diagrammatically below:

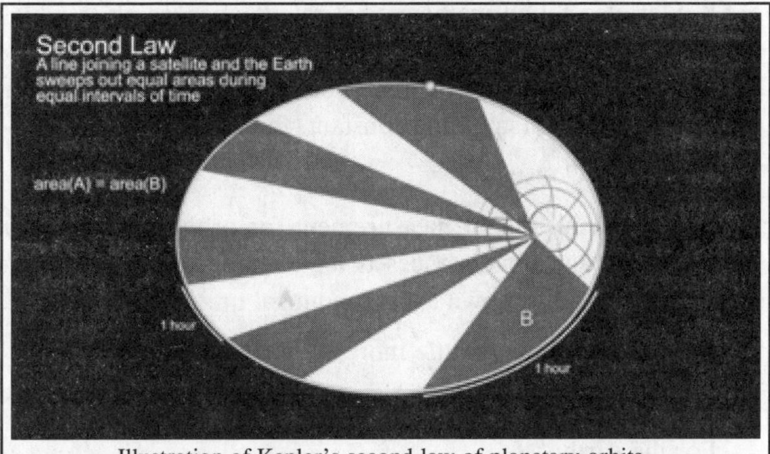

Illustration of Kepler's second law of planetary orbits.
The line joining of the moon to the earth (or a planet to the sun) sweeps out equal areas in equal times as the planet travels around the ellipse.
Credits: NASA Scientific Visualization Studio
Visualizations by Kel Elkins Released on May 7, 2018
https://svs.gsfc.nasa.gov/4642

Thus, the earth executes elliptical motion around the sun with constantly changing angular speed. It moves faster when it is near the sun (known as Perihelion) and slower when it is far from the sun (known as Aphelion)

Kepler's Third Law of Orbital Motion also known as the law of Harmonies

Kepler's first and second laws describe the motion characteristics of a single planet. The third law makes a comparison between different planets. It compares the orbital period and radius of orbit of a planet to those of other planets. The period for a planet to orbit the sun increases rapidly with the radius of its orbit. Thus, Mercury, the innermost planet (closest to the sun), takes only about 88 earth days to orbit the sun, the outermost Pluto (now known as dwarf planet since the year 2006) requires about 248 earth years to do the same. Our earth takes about 365 days (one year) to orbit the sun.

Kepler also found mathematical relation according to which square of orbital period (say T) divided by cube of the distance between two objects, one which is orbiting and the other about which it is orbiting (say R) is constant, that is

T^2/R^3=Constant

A convenient unit of measurement for orbital periods is in "earth years" and for distances is the average separation of the earth from the sun, known as astronomical unit, AU.

Let us discuss this a little more.

In our solar system, all the planets are orbiting the sun.

For the earth:

Orbital period T = 1 earth year

Average distance from sun (orbital radius) R = 1 AU

Therefore, The ratio T^2/R^3 = 1

Let us check this ratio for Mercury and Pluto which are nearest and farthest to sun respectively.

	T in earth Year	R in AU	T^2/R^3
Mercury	0.25	0.3957	1
Pluto	248.08	39.46	1

(Numbers have been rounded off for simplicity)

This can be shown for other planets also.

With the help of Kepler's third law, we can find orbit time of any object orbiting another object at a known orbit radius.

Q196. What is the Tsiolkovsky's rocket equation?

A196. Rockets are momentum (product of mass and velocity) machines. They continuously expel large quantities of burned fuel gases forcibly out of a nozzle at high velocity. These burned fuel gases (exhaust) have both mass and velocity, and therefore some momentum. By conservation of momentum, the rocket's momentum changes by this same amount (with the opposite sign) and it moves in the opposite direction (Newton's third law of motion). Rocket equation is based on Conservation of momentum. This was first done by Russian scientist Konstantin Tsiolkovsky in 1903. All rockets are governed by Tsiolkovsky's rocket equation.

The rocket equation contains three variables.

1. The energy required against gravity (often called "delta V") for reaching the destination, for example from the surface of earth to earth-orbit, earth-orbit to surface of the moon and so on.

"Delta v" is how fast you want the rocket to go. More specifically, it is the change in speed the rocket can achieve.

You can also think of the "delta-v" as proportional to the amount of fuel needed.

2. The energy available in a rocket propellant (often called exhaust velocity "Ve" or specific impulse). "Ve" is the exhaust velocity and represents the total force that the rocket can create. This is used to achieve the delta-v required for a mission.

There are many different types of rocket engines using different fuels. We are most familiar with chemical rockets which use solid or liquid fuel but there are also nuclear and ion rockets. We need a way to compare the amount of force that they produce. This is done by looking at their "Ve". A rocket is generally more efficient if it has better exhaust velocities assuming the total mass of the rocket is the same.

Specific impulse defines how effectively a rocket uses propellant.

3. The propellant mass fraction, shown as "m0/mf" is the total rocket mass for a given destination divided by the dry rocket mass (i.e., without the propellant). "m0" is the initial mass of the rocket and "mf" is final mass after part of its flight.

The mass ratio directly depends on the delta-v and exhaust velocity

The mass fraction tells us what per cent of the total mass of the rocket will be fuel? We would like an engine which will require the smallest percentage of propellent so that we can carry more payload. For example, if 80 per cent of a spaceship needs to be fuel, that leaves 20 per cent free to carry other stuff.

How Tsiolkovsky's rocket equation is derived?

The diagram shown below is a simple model of the Tsiolkovsky's rocket equation. Assume a thought experiment where a rocket is already in motion burning fuel whizzing out exhaust gases from the back of the rocket. Since the exhaust has mass and velocity, it has some momentum. Assume this as a "negative momentum". The "positive momentum" of the rocket will move the rocket in the opposite direction of the exhaust gases (Newton's third law of motion).

At some time, the velocity of the rocket is "V", call this as time "t=0". At this stage its mass is "m+delta m" (Figure at left).

After time "delta t", its velocity becomes "v+delta v" and exhaust of mass "delta m" comes out at velocity "Ve" leaving rocket mass as "m" (figure at right).

Mathematical (calculus) analysis of this will result Tsiolkovsky's rocket equation. (Mathematical analysis is not given in this book)

Note: Calculus is a branch of mathematics concerned with the calculation of instantaneous rates of change (differential calculus) and the summation of infinitely many small factors to determine some whole (integral calculus).

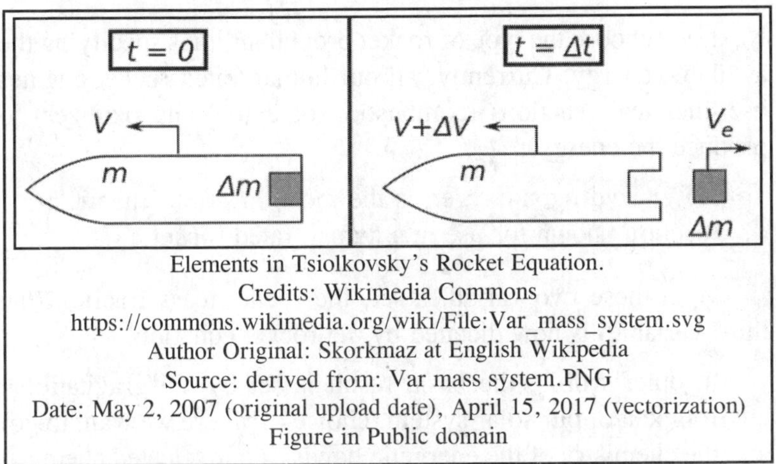

Elements in Tsiolkovsky's Rocket Equation.
Credits: Wikimedia Commons
https://commons.wikimedia.org/wiki/File:Var_mass_system.svg
Author Original: Skorkmaz at English Wikipedia
Source: derived from: Var mass system.PNG
Date: May 2, 2007 (original upload date), April 15, 2017 (vectorization)
Figure in Public domain

Explanation

V is the velocity of the rocket at time t=0 (it will be zero at the start)

m+Delta m is the mass of the rocket at time t=0

V+Delta V is the velocity of the rocket at time t=Delta t

Ve is the velocity of the mass added to the exhaust (and lost by the rocket) during time Delta t m is the mass of the rocket at time Delta t

Q197. How is the Tsiolkovsky's rocket equation used?

A197. The rocket equation allows scientists and engineers to quantify and compare the energy required to reach various destinations in space. It can be applied to orbital maneuvers to determine how much propellant is needed to change to a particular new orbit, or to find the new orbit as the result of a particular propellant burn.

Step 1:

We first select where we want to go. This will specify the energy requirement against gravity for the journey.

Step 2:

Next, we choose the type of rocket propellant, thus specifying the available energy. Currently, all our human rated rocket engines use chemical reactions (combustion of a fuel and oxidizer) to produce the energy.

> **Note:** Hydrogen-oxygen is the most energetic chemical reaction known for use in a human rated rocket.

With these two variables set, the rocket mass fraction (the third variable) is now dictated by the rocket equation.

In other words, the basic result is set by the gravitational environment of our solar system (choice of where we want to go) and the chemistry of the energetic bonds of our selected chemical components (choice of propellant).

Q198. What conclusions can be drawn from the Tsiolkovsky's rocket equation?

A198. Without going into the mathematics of the rocket equation, the following conclusions can be derived:

1. Energy required for travelling from the surface of the earth to earth-orbit (that is escaping out of earth's gravity) is one of the most energy intensive steps of going anywhere else. It is due to the magnitude of earth's gravity and near zero gravity in space.

   For example, if we divide going to the moon from earth's surface in two parts:

(a) From earth's surface to earth-orbit, say about 400 km (250 mi)

(b) From earth orbit to the moon about 385000 – 400 = 384600 km (239000 mi)

The Rocket Equation tells that the energy required for the second step, a journey of 384600 km (239000 mi), is only 70 per cent of the first step which is a journey of only 400 km (250 mi).

Some scientists put it as follows:

It takes more than double the delta-v (representation of energy requirement) to reach Low earth Orbit (LEO) from earth's surface than it takes to reach Low Lunar Orbit (LLO) from LEO.

All destinations between the LEO and the moon are but a fraction of that required to just reach LEO from the earth's surface.

This is remarkable as it means that the first barrier to space (earth's surface to LEO) is much higher than the subsequent ones

The huge cost of this first step is due to the magnitude of earth's gravity. Some rocket scientists say that the giant leap for mankind is not the first step on the moon, but in attaining earth-orbit.

2. The satellite is only a small fraction of the total mass of the rocket that lifts it and yet has an effect on the rocket itself. As a rule of thumb, fuel requirements increase exponentially with every step increase in mass added to the satellite. That is why rockets are mostly fuel and only some metal. As an example, Saturn V that was used in Apollo11 was 85 per cent fuel, 13 per cent rocket and the rest 2 per cent was the moon bound spacecraft with astronauts sitting inside.

3. If the earth was more massive, a rocket would have to expend more energy against a more gruesome gravity well. More fuel would be required to get to space. Increase earth's mass further and the fuel-to-mass ratio would start increase very

steeply and a point will come where it's simply not possible to engineer such a near-all-fuel rocket. It turns out that if the earth was 50 per cent more massive, we would not be able to venture into space, at least using rockets for transport even with the most energetic fuel combination—liquid hydrogen and liquid oxygen—available in chemical rockets.

4. The moon has a much weaker gravity than earth thereby putting the consequences of the rocket equation in our favor and allowing rockets to take off from its surface with ease, almost six times easier. Some scientists have estimated that it takes more than 10 times more energy, theoretically, to get into geosynchronous earth orbit (GSO) from the surface of the earth than from the surface of the moon.

5. With the help of rocket equation, we can calculate the mass fractions of fuel and rocket (including pay load) for earth orbit. As an example, if we use Hydrogen-Oxygen as propellent, the fuel mass will be 83 per cent (For Kerosene-Oxygen propellent it will be 94 per cent) and rocket plus payload will be only 17 per cent. This tell us how little payload is permitted. The Saturn V payload (Apollo 11) to earth orbit was about 4 per cent of its total mass at liftoff.

Note:

Low-earth orbit (LEO) encompasses earth-centered orbits with an altitude of 2,000 km (1,200 mi) or less. For the purposes of the Commercial Use Policy, low-earth orbit is considered the area in earth orbit near enough to earth for convenient transportation, communication, observation and resupply. This is the area where the International Space Station currently orbits and where many proposed future platforms will be located.

Low-Lunar Orbit (LLO): Lunar orbit is the orbit of an object around the moon. LLO are orbits below 100 km (62 mi) altitude.

A geosynchronous orbit (GSO) is an earth-centred orbit with an orbital period that matches earth's rotation on its axis. A circular geosynchronous orbit has a constant altitude of 35,786 km (22,236 mi).

A gravity well is the pull of gravity that a large body in space exerts. The larger the mass of the body the more of a gravity well it has)

There is an advantage to using hydrogen-oxygen as a high-performance propellant; however, it is technically more complex. Kerosene offers less performance but gives a simpler, robust, and easier to fabricate rocket.

Q199. What is colloquially known as the "tyranny of the rocket equation"?

A199. Tyranny is a human trait that we sometimes project onto Nature. This projection is a form of rationalization, perhaps a means to cope with matters that we cannot control. Such is the case when we invent machines to free us from the bounds of earth, affecting our escape into space. If we want to expand into the solar system, this tyranny must somehow be deposed.

Tsiolkovsky's rocket equation tells us that we are distinctly disadvantaged in seeking to achieve orbit. The problem known as the "tyranny of the rocket equation," is that as payload mass increases, so does the amount of propellant required to break free of earth's gravitational grip.

Scientists believe that if we want to break the tyranny of the rocket equation, new paradigms of operating and new technology will be needed. One of the possibilities is to learn how to use raw materials from sources outside the earth. Fortunately, metals key for industrial application, such as aluminum, nickel, cobalt and titanium, happen to be incredibly abundant throughout the inner solar system. With its gravity well only one sixth of earths and a rich array of chemical elements, the moon will be a staging ground for future missions. However, for that to occur, early human explorers will need significant materiel support. That is where NASA's Space Launch System (SLS), a new giant moon rocket, will make all the difference. The rocket is to form the backbone of the space agency's plans to return astronauts to the moon for the first time in decades within the next few years, and onwards to Mars. SLS is a super heavy lift launch vehicle that provides the foundation for human exploration beyond earth's orbit. With the evolution of the SLS to its Block 2 variant with a payload capacity of nearly 100,000 pounds per launch, the nation will finally realize the dream of a permanent lunar outpost.

Another possibility to break the tyranny is the discovery of some new physical principle which could break the tyranny and allow earth escape outside the governance of the rocket paradigm.

References

1. *Mehta, Jatan. Rocket equation 101: The moon as a launch platform. Jatan Space. June 11, 2021* https://blog.jatan.space/p/the-moon-as-a-rocket-platform?s=r
2. *Pettit, Don (NASA Astronaut). The Tyranny of the Rocket Equation. Nasa.gov feature. Last Updated: May 1, 2012. Page Editor: Amiko Kauderer, NASA Official: Brian Dunbar* https://www.nasa.gov/mission_pages/station/expeditions/expedition30/tryanny.html
3. *Mehta, Jatan. Rocket Science 101: The tyranny of the rocket equation. Team Indus Blog. May 24, 2018.* https://medium.com/teamindus/rocket-science-101-the-tyranny-of-the-rocket-equation-491e0cf4dc6a
4. *Zur, Christian. Escaping the Tyranny of the Rocket Equation. Scientific American. February 18, 2020.* https://blogs.scientificamerican.com/observations/escaping-the-tyranny-of-the-rocket-equation/

Q200. How are the science laws used in space travel?

A200. Kepler's Laws of Orbital Motion and Newton's Laws established the basic principles governing the propulsion and orbital motion of modern spacecraft. Tsiolkovsky rocket equation is used to calculate quantity of propellent required.

Any space travel requires:

1. Take off (launching) of a spacecraft,
2. Taking it to a predetermined orbit,
3. Manouevring it in space

Let us see how the various laws of science mentioned above are used in doing these tasks.

1. Take off (Launching) of Spacecraft

Space rockets, used for launching spacecraft, work on Newton's third law of motion (For every action there is an equal and opposite reaction) and use Tsiolkovsky rocket equation to calculate quantity of propellent required for a particular mass of rocket, pay load and rocket. To understand this, take a balloon filled with air. Hold the balloon at its mouth pointing downward. Now leave the balloon. As the air will rush out downward, the balloon will go upward until all the air is out of the balloon. If somehow you let the air continuously coming out, the balloon will keep on going upward till the air stops coming out of the balloon. Coming out of air is "action" while movement of balloon is "an equal and opposite reaction". In the case of rockets, the massive force generated by hot gases from combustion of propellent firing backward from a rocket's engine (the action) produces an equal and opposite force (the reaction) that pushes the rocket forward through space.

2. Reaching into an Orbit

The trick to launch a spacecraft is getting high enough to do its job without losing it to outer space. It's a delicate balance of push and pull, accomplished by the inertia of the moving object and the earth's gravity. By making calculations it can be shown that if a spacecraft is launched at 27,200 km per hour (17,000 miles per hour) the forward momentum will balance gravity and it will circle the earth. On the other hand, if the spacecraft is launched faster than 37,600 km/hr (23,500 m/hr) it will leave the gravitational pull of the earth.

3. Manoeuvring in Space

There is no ground, no water and, no air in space against which a spacecraft can react to manoeuvre. Manoeuvring a spacecraft in space, therefore, is unlike maneuvering any other vehicle on ground, water, or air. Every manoeuvre of a spacecraft in space is achieved by firing a rocket engine and throwing burn fuel using Tsiolkovsky rocket equation. The spacecraft reacts by moving in the opposite direction as per Newton's third law of motion. The motion of the spacecraft is fully determined by gravitational forces acting on the spacecraft when the rocket burn is completed. After the burn any changes in the speed or direction of the spacecraft are beyond the control of the crew. Gravity rules. Rocket engines can be fired in the same direction of the orbit or in the opposite direction or at any angle depending upon the objective of maneuver.

In the case of a spacecraft to land on the moon, it can be taken in the gravitational field of the moon by proper maneuvering and can be landed on the surface of the moon at a predetermined site. Take off from the moon is like that of the earth except it takes much less fuel due to lower gravity on the moon.

Q201. How do satellites stay in orbit?

A201. Satellite Motion

Have you ever thought that if gravity pulls satellite toward earth, why doesn't it fall as a ball thrown in air would? The answer is that satellites do not travel straight up into the air, they move around the earth. If we throw a ball horizontally in front of us, it will move out in front of us and then fall on the ground. If we through the ball faster, it will land even farther in front of us. The faster we throw, the farther it travels before it lands. Newton wondered what would happen if we were on a high mountain and were able to throw an object as fast as we wanted. The faster we throw it, the farther away it would land. At a certain speed, the path of the object would match the curve of the earth. Although the object would keep falling due to gravity, earth's surface would curve away from the object at the same rate. Thus, the object would circle the earth. This is the principle used in satellites. Satellites in orbit around the earth continually fall toward earth, but because earth is curved, they travel around it. Therefore, a satellite falls around earth rather than into it. A satellite does not need fuel because it continues to move ahead due to its inertia (Newton's first law). At the same time, gravity continuously changes the satellites direction.

Why does a spacecraft stay in an orbit?

As per Newton's first law of motion, whenever a spacecraft is launched into space, it tries to keep on moving in a straight line in the direction of force applied to it. But the gravitational pull of the earth tries to bring it back to the earth. If a balance can be maintained between these two opposite forces, then the spacecraft will stay in an equilibrium stage. It neither will go up nor will come down.

As mentioned earlier if the launching velocity of the spacecraft is 27,200 km per hour (17000 miles per hour) then the spacecraft will not come back to earth and will start orbiting in an orbit as per Kepler's laws of orbital motion. This velocity is sufficient for a spacecraft to shut off its engines and continually orbit the earth like the moon.

The orbital velocity of a spacecraft depends on its altitude above earth (Kepler's third law). Nearer to earth, faster is the required velocity because the gravity's pull is large. Orbital velocity at any altitude is calculated by Kepler's third law of orbital motion. For example, at an altitude of 200 km (125 miles) orbital velocity is about 27,400 km per hour (17,125 miles per hour). At moon's altitude, it is 3,680 km per hour (2,300 miles per hour).

Geostationary Orbit (GEO)

In this reference it is of interest to mention about an orbit in which a spacecraft takes 24 hours to complete one revolution of the earth which is equal to the earth's rotational period around its axis. Any spacecraft in this orbit appears to be motionless in the sky with reference to the earth. This orbit is known as Geostationary or Geosynchronous orbit. By Kepler's third law it can be calculated that such an orbit is about 35,200 km (22,000 miles) above sea level. GEO is of great interest to operators of communications and weather satellites. The concept of GEO satellite for communication was first published in 1928 by Herman Potocnik, an Austrian rocket engineer. However, it was first published on a wider scale in 1945 by Sir Arthur C. Clarke, a science fiction author and inventor (a British and Sri Lankan national). As a result, this is sometimes referred to as Clarke Orbit. The Clarke orbit is about 264,000 km (165,000 miles) long.

❑

CHAPTER 17

Space Race Between USA and USSR (Soviet Union)

Q202. What started the space race between the USA and the Soviet Union (USSR) in 1950s?

A202. There is a deep-rooted history to why do United States and Soviet Union compete. In the 1930s Germany began research into the development of rockets for military use. After the German defeat, the World War II (WW II) ended splitting the world into two major camps United States (US) and Soviet Union (USSR). Following the Second World War, many German rocket scientists went to the United States and a lesser number went to the Soviet Union. Tensions rose between two of the superpowers. The primary source of conflict arose from the struggle between two political beliefs of communism (USSR) and democracy (US). This conflict became known as the Cold War. The term cold in this context meant that there was no direct fighting between the two sides, but each sponsored and supported many conflicts across the world. Each side developed powerful and deadly weapons and tried to dominate militarily. Both developed hydrogen bombs in the 1950s. That mounted tensions between the two countries. By the mid-1950s, the Soviet

Union was years ahead of the United States in terms of rocket technology. In 1955, both countries started building ballistic missiles that could be used to launch objects into space. This was the beginning of the space race.

Q203. What were some of the initial events of the space race between the USA and USSR? (Up to the first Soviet and the first American orbited the earth)

A203. In July 1955, the United States announced plans to launch an earth orbiting satellite. Four days later the Soviet Union also announced their intention to launch a satellite soon. The Soviet Union wanted to place a satellite in earth's orbit before the United States to show the world the superiority of Soviet's technology and the communist system. They began modifying their long-range military rockets for use as a satellite launch vehicle. The United States, probably, did not know about the development taking place in the Soviet Union regarding sending a spacecraft in earth's orbit. Some of the initial events of the space race up to the first Soviet and the first American orbited the earth were:

1. Soviet Union launched an earth satellite into space (October 4, 1957)

The Soviet Union launched Sputnik 1 on October 5, 1957, the world's first artificial satellite into space. Sputnik 1 took about 98 minutes to orbit the earth. This indicated that the Soviet Union could use ballistic missiles carrying hydrogen bombs and can attack any target on earth. The United States did not have this capability. This frightened the Americans.

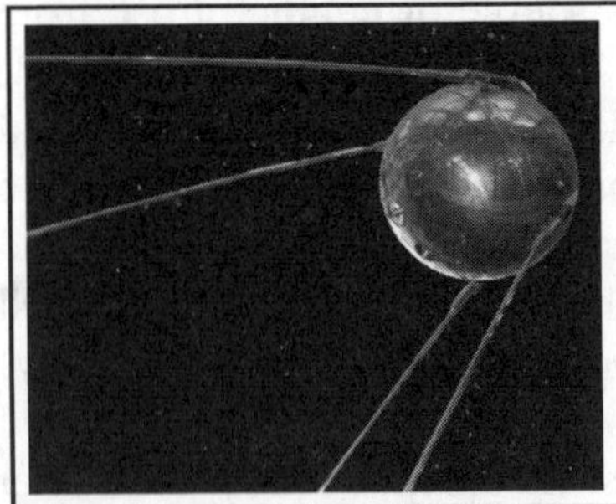

A replica of Sputnik 1, the first artificial satellite in the world to be put into outer space:
Source: Wikimedia Commons
https://en.wikipedia.org/wiki/Space_Race#/media/File:Sputnik_asm.jpg
Credits: NSSDC, NASA
Date 2004
Figure in the public domain

2. Soviet Union launched another satellite carrying a dog (November 3, 1957)

The Soviet Union surprised the world again and launched Sputnik 2 on November 3, 1957, carrying a dog named Laika. The Soviets demonstrated their technological superiority.

3. United States launched a satellite (January 31, 1958)

The United States launched Explorer 1 on January 31, 1958. It took about 1148 minutes to orbit the earth. The US constituted National Aeronautics Space Administration (NASA) on July 29, 1958, to expedite the programme.

Now the cold war had established as a full-fledged space race.

4. Soviet rocket hit the moon After 35 hours (September 13, 1959)

A Soviet space probe, Luna 2, was launched on September 12, 1959, and impacted the moon on September 13, 1959. It was the first spacecraft to reach the surface of the moon, and the first human-made object to contact another celestial body.

5. Soviet orbited Cosmonaut Yuri Gagarin around earth and recovers him (April 12, 1961)

Both the Soviet Union and the United States raced to get a human into space. The Soviet Union sent Major Yuri A. Gagarin into space in the first human spaceflight. He orbited the earth for 108 minutes completing one orbit of the earth on April 12, 1961. The spacecraft made its reentry over the Soviet Union, with Gagarin ejecting from the spacecraft and landing by parachute.

6. US sent man 115 miles into space (May 5, 1961)

On May 5, 1961, the United States launched a Mercury spacecraft into space performing a suborbital flight piloted by astronaut Alan Shepard. He became the first American in space. The flight lasted less than 16 minutes and attained an altitude of just over 187 kilometres (116 miles). Shepard did not orbit the earth, but simply went up and down. He did, however, become the first astronaut to safely return to earth inside his vehicle.

By 1961, some thought that the Soviet Union clearly had superior rocket technology.

7. Astronaut John Glenn of USA orbits earth 3 times (February 21, 1962)

On February 21, 1962, John Glenn became the first American to orbit the earth. He completed three orbits of the earth. After 4 hours and 56 minutes in flight the spacecraft re-entered the earth's atmosphere, splashed down in the Atlantic Ocean.

The world could now see that the United States could keep up with the Soviet's space technology.

This went on. Both the superpowers were active in the space race. Both took a challenge to send men on the moon.

Q204. How did the space race between the USA and Soviet Union benefit the world?

A204. Was there any winner of the space race between the USA and the Soviet Union? The important point is not who, if any, won the race but the space race yielded considerable benefits for human society. It produced groundbreaking efforts to launch artificial satellites, space probes to the moon, Venus, and Mars, and human space voyages in low earth orbit and lunar missions. It produced rapid improvements and advances in many fields such as telecommunications, micro-technology, computer science, solar power, health and medicine, transportation, public safety, consumer goods, energy and environment, information technology, and industrial productivity. Here are some of the ways the world has changed.

Communication

We now have a network of satellites which provide broadband communications and high-definition television, data used for weather reporting and GPS navigation and positioning. Many of these tools and systems were created and developed during the space race.

Medical

Technology developed for deep space imaging and photography is being used in CAT scans and radiography Shock absorbent materials developed by NASA are helping create more functionally dynamic artificial limbs.

Technology

Many inventions we use every day were created for space exploration. Some of the well-known technologies/products developed during space exploration activities and adapted for the commercial markets are memory foam (originally named temper foam), freeze-dried food, firefighting equipment, Dust Busters, Artificial limbs, Wireless headsets, CAT scans, Portable computer, Computer mouse

❑

CHAPTER 18

Journey to the Moon and Lunar Exploration

Q205. Who discovered/studied the moon before the Apollo mission?

A205. The list of scientists/astronomers, starting from ancient Greek philosophers, who discovered/studied the moon before humans arrived on the moon is very big. These scientists made some fundamental discoveries about the moon. Some of the important names and their contribution towards knowledge about the moon are given below.

1. The ancient Greek philosophers

Anaxagoras (d. 428 BC) reasoned that the sun and moon were both giant spherical rocks, and that the moon reflected the light of the sun.

Plutarch (AD 46 – after AD 119) suggested that the moon had deep recesses in which the light of the sun did not reach and that the spots are nothing but the shadows of rivers or deep chasms. He also suggested the possibility that the moon was inhabited.

Aristarchus of Samos (c. 310 – c. 230 BC) computed the distance of the moon from earth, together with its size, obtaining a value of 20 times the earth radius for the distance (the real value is 60; the earth radius was roughly known since Eratosthenes (c. 276 BC – c. 195/194 BC).

2. **Chinese.** The Chinese of the Han dynasty (202 BC–202 AD) recognized that the light of the moon was merely a reflection of the sun. This was supported by thinkers such as Jing Fang, who noted the sphericity of the moon. Shen Kuo (1031–1095) of the Song dynasty (960–1279) created an allegory equating the waxing and waning of the moon to a round ball of reflective silver that, when doused with white powder and viewed from the side, would appear to be a crescent.

3. **Indian:** By 499AD, the Indian astronomer Aryabhata mentioned in his Aryabhatiya that reflected sunlight is what causes the moon to shine.

4. **Islamic Scholars:** Habash al-Hasib al-Marwazi, a Persian astronomer, conducted various observations in Baghdad between 825 and 835 AD. He estimated the moon's diameter as 3,037 km (1887 mi) (actual 2160 mi) and its distance from the earth as 346,345 km (215,209 mi) (actual 238 900 mi). In the eleventh century, the Islamic physicist Alhazen investigated moonlight and concluded that it was a combination of the moon's own light and the moon's ability to absorb and emit sunlight.

5. Galileo Galilei, Riccioli, and Grimaldi

Galileo was the first scientist who made his conclusions based on observations rather than speculations. He recognized that the moon was not the perfect sphere as earlier thought. In 1609, he drew one of the first telescopic drawings of the moon and noted that it was not smooth but had mountains, valleys, and craters. He studied lunar topography, recognized that the dark areas (which he named maria (Latin for "seas") were lower and smoother than the brighter regions. He also concluded that if it had any life, it would have to be different from that on earth.

Later in the seventeenth century, Giovanni Battista Riccioli (April 17, 1598 - June 25, 1671) and Francesco Maria Grimaldi (April 2, 1618 - December 28, 1663) drew a map of the moon and gave many craters the names they still have today.

6. Van Langren and Johannes Hevelius

The first map of the moon was made by the Belgian astronomer Michael Florent van Langren in 1645. Two years later, in 1647, The Polish astronomer Johannes Hevelius published the book Selenographia, containing 40 engravings of the moon, including drawings of waxing and waning phases, and three full maps. It was the first book to describe the moon.

7. Adrien Auzout

In 1666, the French observer Adrien Auzout argued that the moon was not inhabited as earlier thought. He also made observations of comets and argued in favor of their following elliptical or parabolic orbits

8. Johann Madler and Wilhelm Beer

In 1834–1836, German partners, Wilhelm Beer and Johann Heinrich Madler published the first exact map of the moon, published in four volumes and a book in 1837, which firmly established the conclusion that the moon has no bodies of water nor any appreciable atmosphere. Beer and Madler's book and maps became the standard reference for nearly 100 years.

9. G. K. Gilbert

In 1893, the American geologist Grove Karl Gilbert pointed out that volcanic craters were typically at the tops of massive mountains, whereas lunar craters usually occur on flat terrain. He concluded that impacts had formed most craters. He suggested that the basin containing Mare Imbrium was the moon's largest impact crater.

10. Ernst Opik and A.C. Gifford

Until the early 1900s, nearly all observers had accepted that lunar craters were volcanoes because impact craters were unknown on earth. However, in 1916 and 1924, an Estonian physicist Ernst Opik, and the New Zealander A. Charles Gifford independently provided evidence that craters on the moon are the result of meteorite impact.

11. Robert Dietz

In 1946 Robert Dietz, an American geologist recognized that crater morphology changed with the energy of the impactor, as measured by crater diameter. He also noted that maria are found within the largest impact craters and that later impacts degraded existing landforms. He also said that the moon would now be cold and dead.

12. Ralph Baldwin

In the 1940s, the American planetary scientist Ralph Baldwin realized that extraordinary energy would have propelled mountains of rock hundreds of kilometres beyond Imbrium's rim. He proved that the depths of craters formed via explosions were proportional to their diameters,

13. Eugene Shoemaker

In the late 1950s and early 1960s, the geologist Eugene Shoemaker studied impact mechanics, including the strings of small bright craters and bright rays. According to him, these were due to debris ejected from high velocity impact events. With fellow geologist R.J. Hackmann, Shoemaker compiled a geologic map of Copernicus and the adjacent Mare Imbrium, establishing a system of lunar stratigraphy, which was later used across the solar system.

14. William K. Hartmann

Working with Kuiper, Hartmann recognized that the small Mare Orientale on the western limb was the centre of a large, three-ring structure that he called a multi-ring basin. Hartmann and Kuiper documented similar concentric and radial structures around every nearside circular mare. All of these turned out to be lava-filled impact basins. Scientists accepted that impacts had formed lunar craters, so an older surface would have more craters than a young surface. Hartmann determined age of maria as 3.6 billion years, exactly the age of Apollo 11's samples from Mare Tranquilities.

References

1. *Wikipedia. Exploration of the Moon.* https://en.wikipedia.org/wiki/Exploration_of_the_moon
2. *Wood, Charles A. Before Apollo: The Scientists who discovered the Moon. Sky and telescope. July 9, 2019* https://skyandtelescope.org/astronomy-resources/before-apollo-scientists-who-discovered-moon/

Q206. Why is space exploration important?

A206. Space exploration is important not only to humanity's curiosity, but it is also important for the future of the earth. It provides benefits in many areas such as health, mining, security, and inspirational. Here are some of the arguments in favor of continuing space exploration:

1. According to a 2012 survey of research, most scientists estimate that earth has a carrying capacity of between 8 and 16 billion people. We already have a population of over 7 billion. We therefore should plan to colonize other places in space like the moon, Mars, etc.
2. Being overpopulated on earth, we are consuming the earth's natural resources quickly. Space is believed to hide valuable substances. The moon has helium-3 used for certain medical applications and a possible fuel for nuclear power plants. The moon also has some rare earth elements such as europium and tantalum useful in electronics and solar panels. Mars is one of the largest exposures of clay minerals. There is gold, silver, platinum, and other valuable substances in the cosmos.
3. We need to protect ourselves against the threat of being hit by asteroids. Space programme will enable us to spot a dangerous object long before it strikes earth, and we can device methods to change its collision course.

4. Space exploration helps humans develop new technologies, make new innovations, and develop new materials and gadgets. Some of the medical innovations which resulted from space exploration are ear thermometers, automatic insulin pumps, laser angioplasty, voice-controlled wheelchairs, and programmable pacemakers. Freeze-dried food and heat reflecting materials are also from space exploration programme.
5. Space exploration Promote Science Education and related fields. Agencies involved in space exploration studies have done extensive good environmental research in studying air quality, climate change and alternative energy.
6. Space exploration is important for National Security. It will help detecting and preventing a hostile nation or terrorist group from deploying space-based weapons, or attacking its navigational, communications and surveillance satellites.
7. Humans have always been curious to find hidden mysteries of the universe that lie far beyond our earth. Answers to many unanswered questions will be found by exploring space. Through space exploration we can search for the chemical signs of life in the atmospheres of distant planets outside our solar system.

References

1. *Hohler, Daniel. Top 5 Reasons why Space Exploration is Important for the World. PlanetSave. July 26, 2009.* https://planetsave.com/2009/07/26/top-5-reasons-why-space-exploration-is-important-for-the-world/
2. *Kiger, Patrick J. 10 Reasons Why Space Exploration Matters to You. HowStuffWorks/Science/Space/Space Exploration. Updated: April 7, 2021* https://science.howstuffworks.com/10-reasons-space-exploration-matters.htm

Q207. Why do we want to go back to the moon?

A207. The Luna and Apollo missions taught scientists just how valuable the moon could become. Exploration of the moon has thus become a global undertaking. Here are some reasons to go back to the moon:

1. The moon has valuable minerals: Our earth does not have infinite resources to sustain life forever. We thus must find some other sources of such material to help life on earth. Analysis of the lunar rock and dust has revealed that the moon contains a plethora of minerals of high industrial value. The discovery of water on the moon and potential deposits of rare minerals hold promise for both scientific and economic exploration and exploitation. New instruments are being developed to test and extract resources like water.

2. The moon has helium-3 for clean and safe nuclear fusion power and medical science: Scientists have discovered plenty of helium-3 in the lunar soil which is extremely rare on earth. Helium-3 could be used in waste free nuclear fusion reactors. Another use of helium 3 is in the production of short-lived medical isotopes used in medical science. It has been estimated that there are around 1.1million tons of helium-3 on the surface of the moon down to a depth of a few metres enough to power human energy needs for up to 10,000 years.

3. Prepare for deep space travel: The moon could be a base for deep space travel. A lunar base could be a site for launching rockets with locally manufactured fuel to distant planets such as Mars. Lower gravity of the moon makes spacecraft launching economically attractive due to lesser fuel consumption. Scientists want to develop and test technologies to pave the way for future deep-space missions.

4. Further exploration of the moon will help in learning more about the history and formation of the universe and our earth.

The moon and earth were both formed during the creation of our Solar System, approximately 4.6 billion years ago. The moon is free from wind and water erosion, so things stay as is for millions of years. The moon has the most ancient surfaces, some up to 4.5 billion years old near the start of the solar system. Here on earth these ancient surfaces are destroyed by erosion and plate tectonics (movements within the earth's crust). So, a record of what the early earth was like is preserved in rock samples from the moon. Studying lunar rocks not only helps us to understand about the moon, but also allows us to understand what the early earth was like. Similarly, by studying the moon's craters, scientists infer what must have happened in the solar system's past. As our nearest neighbour, the moon is a natural laboratory for exploring fundamental questions about the origin and evolution of the earth and the solar system.

5. Radio Astronomy without interference from Earth

A radio telescope on the moon's far side would have many advantages. earth's ionosphere blocks earth radio telescopes from seeing longer wavelengths. The moon's lack of an atmosphere will allow the longer radio wavelengths to reach a telescope built on the moon. The far side of the moon is an excellent site because the moon itself will block the radio chatter emitted from earth. These advantages will open a range of wavelengths that astronomers have not been able to explore so far. This will advance human knowledge of the Cosmos and searching for signals from extraterrestrial civilizations

6. For our long-term survival we must search for other sites in the solar system. Carl Sagan once said, "All civilizations become either spacefaring or extinct." The last few decades of space exploration and astronomy have shown that the universe is violent and dangerous, at least with respect to human life. There is always a danger of some space object hitting our earth. In 1908 an object of unknown nature (probably a comet) hit Siberia with a force equivalent to a hydrogen bomb. 65 million years ago, an impact wiped out the dinosaurs living on earth at the time. Thus, humanity is vulnerable as long as we are confined to one planet. We can increase our chances of long-term survival by dispersal to other sites in the solar system. Where can we go? The moon's proximity to earth is beneficial to learn, to work, and to prepare for more ambitious voyages deeper into space. Now, human life exists only on the earth. But with modern technology, there are several other possibilities, starting with the moon itself. Space exploration organizations worldwide are therefore contemplating to make lunar colonies so that some people can be offloaded from the earth and start living there. That is one of the reasons that this time they are not simply aiming to go to the moon, but rather to go there to stay. They want to investigate the possibility of establishing bases on the moon's surface for a long-term presence there. Other possibilities include putting humans on Mars and beyond.

7. Lunar-based solar power system: Scientists have advocated obtaining solar power from the moon. They are of the opinion that a lunar-based solar power system would be cheaper than earth-based solar power implemented on a global scale. The key is the moon's lack of atmosphere and weather which will help making photovoltaic cells paper thin in contrast to heavy structures needed on earth. This will make collection of solar power on the moon cheaper. Scientists have proposed the large-scale construction of solar collectors on the lunar surface, using local lunar materials. The solar energy would be converted to microwave energy and transmitted to earth. It seems difficult but not impossible.

8. A place to offload industrial pollution

Industrialists are proposing the moon as an ideal home for factories. Since there are minerals and oxygen in lunar rocks/ dust and frozen water in certain locations, large segments of heavy industry can be relocated there, reducing the amount of pollution that is produced on earth.

References

1. *Mehta, Jagan. Why explore the Moon. Planetary Science, Space Missions. April 3, 2020.* https://blog.jatan.space/p/why-explore-the-moon

2. *McKie, Robin. Everyone's going back to the moon. But why? The Observer / The Guardian. 6, July 2019.* https://www.theguardian.com/science/2019/jul/06/everyones-going-to-the-moon-again-apollo-11-50th-aniversary

3. *Ask a Mathematician / Ask a Physicist. What's the point of going to the moon? July 19, 2019.* https://www.askamathematician.com/2019/07/q-whats-the-point-of-going-to-the-moon/#comment-778002

Q208. How many people have been to space to date?

A208. The criteria for determining who has achieved human spaceflight vary. The Federation Aeronautique Internationale (FAI) defines spaceflight as any flight above 100 kilometres (62 mi). According to this definition, as of February 2023, a total of 631 people from 44 countries have gone into space spending in excess of 177 man-years. 622 people reached earth orbit, 24 travelled beyond low earth orbit and 12 walked on the moon.

Notes:

Low Earth Orbit: A low earth orbit (LEO) is an earth-centred orbit near the planet, often specified as having a period of 128 minutes or less (making at least 11.25 orbits per day). Most of the artificial objects in outer space are in LEO, with an altitude never more than about one-third of the radius of earth.

The term LEO region is also used for the area of space below an altitude of 2,000 km (1,200 mi) (about one-third of earth's radius).

All crewed space stations to date have been within LEO. From 1968 to 1972, the Apollo programme's lunar missions sent humans beyond LEO. Since the end of the Apollo programme, no human spaceflights have been beyond LEO.

Reference

1. *Wikipedia. List of space travellers by name. https://en.wikipedia.org/wiki/List_of_space_travelers_by_name*
2. *Wikipedia. List of space travellers by nationality. https://en.wikipedia.org/wiki/List_of_space_travelers_by_nationality*
3. *Astronaut / Cosmonaut Statistics. Astronauts & Cosmonauts. Worldspaceflight.com. April 30, 2023. https://www.worldspaceflight.com/bios/stats.php*

Q209. How many countries have been to the moon (on its surface or in its orbit)?

A209. Only seven nations (or political unions) have been to or orbited the moon so far as part of a successful or semi-successful mission.

1. The US tops the current list. Between 1964 (Ranger 7) and 2018 (TESS flyby), 32 missions were carried out by the United States. Of these missions, 6 were manned missions: Apollo 11, 12, 14, 15, 16 and 17 for a total of 12 astronauts who walked on the Moon. NASA is planning humanity's return to the Moon in 2024 through the Artemis program.

2. The Soviet Union flew 23 such missions between 1959 and 1976 conducting both the Luna and Zond lunar exploration programmes. But to date, it has never landed a cosmonaut on the Moon's surface.

3. China flew its first mission of the Chang'e programme in 2007. Five attempted missions have been successful and one (Chang'e 3) semi-successful. Rover Yutu of Chang'e 3 that landed on the moon in December 2018 stopped moving after around six weeks but maintained radio contact. China counts seven missions to the moon after a Hong Kong satellite, AsiaSat 3, in 1997 missed its earth orbit, which was subsequently adjusted with the help of two lunar fly-bys, marking the first (accidental) commercial mission to the moon. China plans

to send humans to the moon around 2030. The most notable mission was Chang'e 4: the rover/lander that made a historic landing on the far side of the Moon in January 2019, where no humans or robots had ever been before.

4. India's Chandrayaan-1 successfully completed a lunar orbit and impactor mission in 2008. The second spacecraft Chandrayaan-2 entered the moon orbit in September 2019, but a lander with a rover could not do soft landing due to some software glitches.
5. Japan has also completed two missions, orbiter Hiten in 1990 (semi-successful) and combined orbiter and impactor mission Kaguya in 2007. A private Japanese company, ispace. lifted off its spacecraft Hakuto-R Mission 1 on December 11, 2022, which entered lunar orbit on March 23, 2022. The spacecraft tried to soft land on the Moon on April 25, 2023, but could not complete the landing. Ispace is now developing Mission 2 and Mission 3 concurrently.
6. The ESA (EU) sent probe SMART-1 into lunar orbit in 2003, testing solar-powered ion drive technology. SMART-1 was a small probe designed and built by the Swedish Space Corporation. After orbiting the moon for 3 years, the spacecraft then crashed into the surface, ending its 3 years mission.
7. Luxembourg's company LuxSpace sent a small independent probe on the Chinese Chang'e 5-T1 mission, which orbited the moon and returned to earth in 2014.

Israel is the sole country which can look back only on failed missions to the moon. Their single moon mission, carried out by nonprofit SpaceIL, crash-landed on the moon in February 2019.

Between 1976 and 1990, no moon missions were carried out by any nation.

References

1. *Buchholz, Katharina. Can you list the 7 countries that have been to the moon? World Economic Forum (with the collaboration of Statista). September 11, 2019.*

https://www.weforum.org/agenda/2019/09/which-countries-moon

2. *Moon exploration after Apollo. Smithsonian National Air and Space Museum.* https://airandspace.si.edu/explore-and-learn/topics/apollo/apollo-programme/post-apollo/
3. *Urbain, Tom. How many countries have been to the Moon? Starlust. Last updated 8th July 2020. https://starlust.org/countries-that-have-been-to-the-moon/*

Q210. Why have humans never been back to the moon after 1972?

A210. While more than 100 robotic spacecraft have been launched to explore the moon's surface, only a dozen humans have set foot on the moon. The last was in 1972.

Main obstacles have been justifying the cost and risk to life. In an era of tension between superpowers and in a race to demonstrate their technological superiority, the US and Soviet Union found a reason to fund space programmes. Today, without this threat, similar programmes would need good reasons to argue. Now many nations are active in not only visiting the moon, but a long-term presence.

Q211. What various spacecraft/vehicles have been used so far for the moon exploration, and what did we learn from them?

A211. Some of the milestone spacecraft used in the moon's explorations are given below.

1. In January 1959, Soviet's Luna 1, flew by the moon at 3,725 miles (5,995 kilometres). It was the first spacecraft to reach the vicinity of the moon. Its scientific equipment revealed that the moon had no magnetic field. It also returned evidence of solar wind.

2. In September 1959, Luna 2 impacted the moon, making it the first spacecraft to contact another solar system body.
3. A third Luna was the first-ever mission to photograph the far side of the moon.
4. In 1962 NASA placed its first spacecraft on the moon—Ranger 4. However, due to a failure in the spacecraft's onboard computer, Ranger 4 was unable to return any scientific data before slamming into the far side of the moon.
5. In 1964, Ranger 7 captured more than 4,000 photos before it smashed onto the surface. Images from all the Ranger missions showed that the moon's surface was rough.
6. In 1966 the Soviet spacecraft Luna 9 became the first vehicle to soft-land safely on the surface. The spacecraft photographed a ground level lunar panorama.
7. Luna 10 launched later that year and became the first spacecraft to successfully orbit the moon. It conducted extensive research in lunar orbit, gathering data on the moon's magnetic field, its radiation belts, and the nature of lunar rocks. It found mass concentrations (called "mascons"), areas of high density below the basins.
8. The seven Surveyor probes (1966-68) were the first NASA craft to perform controlled landings on the moon's surface. Surveyor carried cameras to explore the moon's surface terrain, as well as soil samplers that analyzed the nature of lunar rock and dirt.
9. In 1966 and 1967 NASA launched lunar orbiters that were designed to circle the moon and chart its surface in preparation for future manned landings. In total, five lunar orbiter missions photographed about 99 per cent of the moon's surface.
10. On July 20, 1969, Neil Armstrong and Edwin "Buzz" Aldrin became the first people to reach the moon when their Apollo 11 lunar lander touched down in the Sea of Tranquility. Later missions carried a lunar rover that was driven across the

moon's surface, and saw astronauts spend as long as three days on the moon. Before the Apollo project ended in 1972, five other missions and a dozen men had visited the moon.

11. Luna 24 was the last of the Soviet Union's Luna programme. The mission returned 170.1 g (6.00 oz) of lunar samples to the earth on August 22, 1976.

12. Japan's first lunar probe, Hiten was launched in January 1990. Hiten successfully demonstrated the aerobraking technique. Aerobraking is used when a spacecraft requires a low orbit after arriving at a body with an atmosphere, and it requires less fuel than does the direct use of a rocket engine.

13. In 1994, NASA focused on the moon again. The Clementine mission succeeded in mapping the moon's surface. The objective of the mission was to test sensors and spacecraft components under extended exposure to the space environment and to make scientific observations of the moon and an asteroid.

14. NASA's Lunar Prospector (1999) orbited the moon for almost 19 months to map its surface composition and to look for polar ice. The probe found evidence suggesting water ice at both poles. The prospector also explored the moon's gravitational field and remapped its surface. The mission ended with the spacecraft impacting the lunar surface, creating a dust cloud that was studied from earth.

15. SELENE - Kaguya, was the second Japanese lunar orbiter spacecraft following the Hiten probe. the spacecraft was launched in September 2007. After successfully orbiting the moon for a year and eight months, the orbiter impacted the moon on June 10, 2009.

16. Chandrayaan-1 was the first Indian lunar probe. It was launched by the Indian Space Research Organization (ISRO) in October 2008 and operated until August 2009. The mission included a lunar orbiter and an impactor. The moon Impact Probe hit the moon ejecting sub-surface soil that could be analyzed for the presence of lunar water ice.

17. In 2011, NASA launched the twin GRAIL orbiters to make detailed gravity maps of the moon.

18. The Lunar Atmosphere and Dust Environment Explorer (LADEE) was a NASA lunar exploration and technology demonstration mission. It was launched in September 2013. During its seven-month mission, LADEE orbited around the moon's equator, using its instruments to study the lunar exosphere and dust in the moon's vicinity.

19. Chang'e-3 is China's third robotic lunar probe mission. The objective was to soft-land on the moon's surface and deploy an unmanned Lunar Rover to explore the areas surrounding the landing site. It was launched in December 2013.

20. Chandrayaan-2, the second lunar exploration mission of India was launched in July 2019. It had an orbiter, a lander, and a lunar rover. It was launched in July 2019. The lander failed to do soft landing. The mission's main objectives were to study lunar topography, mineralogy, elemental abundance, the lunar exosphere, and signatures of hydroxyl and water ice

21. China's Chang'e 4 landed on the moon in January 2019. It is the world's first mission to land on the moon's far side with a rover, Yutu-2. The rover is scouting the landed region for clues of the moon's past.

22. China's Chang'e 5 is China's first lunar sample-return mission. It was launched on November 23, 2020, landed on the moon on December 1, 2020. It collected 1,731 g of lunar samples and returned to the earth on December 16, 2020. Chang'e-5 was the first lunar sample-return mission since the Soviet Union's Luna 24 in 1976. The mission made China the third country to return samples from the moon after the United States and the Soviet Union.

23. Artemis I was USA's uncrewed Moon-orbiting mission. It was the first integrated flight test of the Orion spacecraft and Space Launch System (SLS) rocket. It was launched on November 16, 2022. Orion completed one flyby of the Moon on

November 21, 2022, and completed a second flyby of the Moon on December 5, 2022. The Orion spacecraft then returned and reentered the earth's atmosphere with the protection of its heat shield, splashing down in the Pacific Ocean on December 11, 2022. The duration of the mission was 25 days, 10 hours and 53 minutes and distance travelled was 1.3 million miles (2.1 million kilometers)

24. Japanese ispace, inc. ("ispace") has built the HAKUTO-R Mission 1 lunar lander (also known as M1 lander). The M1 lander was expected to deploy a two-wheeled, baseball-sized rover from Japan's JAXA space agency and a four-wheeled Rashid rover made by the United Arab Emirates. It was launched on December 11, 2022. It tried to land on April 25, 2023, but failed to soft land. As a follow up, Ispace are developing Mission 2 and Mission 3 concurrently.

Reference

List of rovers on extraterrestrial bodies. Wikipedia
https://en.wikipedia.org/wiki/List_of_rovers_on_extraterrestrial_bodies

Q212. Which important moon exploration activities are being focused on now by global space agencies?

A212. NASA and other global space community have outlined six common areas of activities for exploration of the moon:

1. Extend human presence to the moon to enable eventual settlement.
2. Pursue scientific activities that address fundamental questions about the history of earth, the solar system, and the universe - and about our place in them.
3. Test technologies, systems, flight operations and exploration techniques to reduce the risks and increase the productivity of future missions to Mars and beyond.

4. Provide a challenging, shared, and peaceful activity that unites nations in pursuit of common objectives.
5. Expand earth's economic sphere and conduct lunar activities with benefits to life on the home planet.
6. Use a vibrant space exploration programme to engage the public, encourage students and help develop the high-tech workforce that will be required to address the challenges of tomorrow.

❑

CHAPTER 19

USA's Lunar Exploration Programme

Q213. What were the stages of USA's lunar exploration programme?

A213. American "moon Mission" was developed in several stages. Each stage was an attempt to build on the learnings from the last. The stages were:

1. Probes (or Robots): NASA began launching probes in the late fifties. The probes produced close-up pictures of the moon's landscape, generated maps, and sent robotic landers.
2. Project Mercury: Project Mercury succeeded in sending men in earth's orbit.
3. Project Gemini: Project Gemini had two people with long duration flights and trained astronauts for performing space walks, etc. and
4. Project Apollo: Project Apollo had a goal of landing man on lunar surface.

Q214. What were the goals of various NASA probes (robots) used in lunar exploration activities?

A214. NASA sent Pioneer, Ranger, Lunar Orbiter, and Surveyor spacecrafts to study the moon between 1958 and 1968. These robotic vehicles provided scientists and engineers with a greater understanding of interplanetary space and lunar geography and played an important role in preparing for a manned landing. These were unmanned robotic spacecraft, designed to take photographs, measure temperature, test atmosphere, and send data to earth.

The first series of Pioneer programme (1958-1960)

The Pioneer programme was a series of United States unmanned space missions that were designed for lunar and planetary exploration. The programmes were in two series: the first programme ran from 1958 to 1960 and the second from 1965 to 1992 (will be discussed later). The earliest missions were attempts to achieve earth's escape velocity to show that it was feasible and study the moon. Pioneer-3 and Pioneer-4 (1958-1959) were used for lunar exploration. Pioneer 3, a lunar flyby, missed the moon due to launcher failure. Pioneer 4, a lunar flyby, was launched on March 3, 1959, achieved earth escape velocity. It was the first US probe to escape from the earth's gravity. It passed within 60,000 km of the moon's surface. Pioneer 4 was the first US spacecraft placed in solar orbit. The spacecraft sent back excellent data about the Van Allen Belts.

Ranger programme (1959-1965)

The Ranger series of probes was of hard landers. The objective was to take close-up images of the moon's surface and transmitting them to earth until the spacecraft were destroyed upon impact. This was the first US effort to launch probes directly toward the moon. Out of a total of 9 Ranger spacecraft launched, 5 failed and 4 were successful. Ranger 7 successfully returned images in July 1964, followed by two more successful missions. The images showed the lunar surface in stunning detail which were helpful in the future human exploration of the moon projects.

Lunar Orbiter programme (1964-1967)

The Lunar Orbiter programme was a series of five uncrewed lunar orbiter missions. The programme was initiated in 1964.

The main purpose of the programme was to locate smooth, level areas on the moon's nearside (the moon's side visible from earth) to confirm the areas as suitable for manned landing sites for the Apollo and Surveyor spacecraft by mapping the moon's surface. They provided the first photographs from lunar orbit and photographed both the moon and earth. The specific objectives were photography, selenodesy (measurement of the moon's surface and its gravitational field), and analysis of the moon's environment. With the completion of the five Lunar Orbiter missions, 99 per cent of the moon was photographed.

Surveyor programme (1966-1968)

In NASA's Surveyor programme seven robotic spacecraft were sent to the moon's surface. Its primary goal was to demonstrate the feasibility of soft landings on the moon. Out of a total of seven Surveyor spacecraft, five were successful. Surveyors 2 and 4 failed. All seven spacecraft are still on the moon. Some parts of Surveyor 3 were returned to earth by the crew of Apollo 12. Various sites in the moon's equatorial belt and the southern highlands were examined. Four spacecraft survived the extreme cold of the lunar nights. Besides transmitted over 85000 pictures to earth, chemical and mechanical properties of lunar materials of the surface and near-surface were made. Wide variety of other data were also obtained that greatly increased our knowledge of the moon.

Q215. What was NASA's project Mercury?

A215. Project Mercury (1959-1963) was announced in December 1958 with an objective of sending men in orbit around the earth. The name "Mercury" comes from Mercury, a Roman mythological God who is often a symbol of speed. NASA announced seven astronauts, known as "Mercury Seven", in April 1959. In September 1962, additional nine astronauts were recruited, known as "Next Nine".

The Mercury spacecraft was a cone shaped small capsule with room for one astronaut with a cylinder mounted on top. The astronaut stayed in his seat during the flight.

Astronauts made a total of six spaceflights during Project Mercury. Two of those flights reached space and came right back down. These are called suborbital flights. The other four went into orbit and circled earth.

Some of the highlights of Mercury project are:

- Two chimpanzees, Ham (in Mercury 2) and Enos (in Mercury 5), flew in Mercury capsules. Before these on December 4, 1959, a rhesus monkey, Sam, flew close to the edge of space in the Little Joe 2, a test of the Mercury space capsule, Sam was sent to test the space equipment and the adverse effects of space on humans.
- Alan B. Shepard Jr. was the first American to make a suborbital flight in Mercury 3.
- John H. Glen Jr. was the first American to orbit earth (3 orbits) in Mercury 6.
- L. Gordon Cooper Jr. made 22 earth orbits, stayed in space for more than a day and took a seven-hour nap in space.

Project mercury succeeded in placing six people into space. NASA learned how to put astronauts in orbit around earth, how people could live and work in space, and how to operate a spacecraft in orbit. The project demonstrated that man could function ably as a pilot-engineer-experimenter without undesirable reactions or deteriorations of normal body functions for periods up to 34 hours of weightless flights. These lessons were very important. NASA used them in later space programmes.

Q216. What was NASA's project Gemini?

A216. Project Gemini (1964-1966) was NASA's human spaceflight programme. The Mercury missions had proved astronauts could fly in space. But Mercury flights were of short durations. Astronauts had to prove that they could take space walks and perform repairs in weightlessness. Astronauts had to learn and practice bringing two spacecraft together (Rendezvous) and dock them. That's where project Gemini came in. Each

Gemini spacecraft held two people hence the name Gemini which means Twins in Latin.

The Gemini was a bell-shaped spacecraft. It had three modules: Re-entry module, Retrograde module (moving in opposite direction with reference to a primary object), and Equipment module. It had several new features such as the use of fuel cells instead of batteries, better computers, provision of ejection seats to push astronauts to safety. Spacecraft doors could be opened from inside allowing astronauts to come out in space to take space walk and to do maintenance work.

Project Gemini had 12 flights. The first two were unmanned to test various systems and sub-systems. The first flight to carry astronauts was Gemini III. That flight tested the new vehicle.

Astronauts accomplished many things in the Gemini missions, such as:

- The first US spacewalk (Gemini IV).
- Longer stay in orbit - for more than a week (Gemini V), for two weeks (Gemini VII)
- First rendezvous in space - Gemini VI A and VII missions were in space at the same time and met each other in orbit.
- First docking in space (Gemini VIII)
- Testing different ways of flying near another spacecraft. First spacewalk (Gemini IX)
- Connecting with another spacecraft and using its engines to move both vehicles (Gemini X).
- Flying higher than any NASA mission before (record altitude 1386 km (855 mi) (Gemini XI).

Q217. What was NASA's project Apollo?

A217. Project Apollo (1961-1972) was the NASA programme designed to land humans on the moon and bring them safely back to earth. The mission's goals also included:

- Establishing the technology to meet other national interests in space.
- Carrying out a programme of scientific exploration of the moon.
- Developing human capability to work in the lunar environment.
- Achieving preeminence in space for the United States.

The project had a total of 11 spaceflights. The first four flights tested the equipment used in the programme. The first crewed Apollo flight happened in 1968. The first moon landing took place in 1969 (Apollo 11) and the last in 1972 (Apollo 17). Six of the flights (Apollos 11, 12, 14, 15, 16, and 17) achieved the goal of sending humans to the moon and bringing them back safely to earth. A total of 12 astronauts walked on the moon. They returned a wealth of scientific data and 382 kg (842 pounds) of lunar samples. Experiments included soil mechanics, meteoroids, seismic, magnetic fields, and solar wind. Apollo 1, initially designated AS-204, was the first crewed mission, but it never flew. A cabin fire during a launch rehearsal test on January 27 killed all three crew members. Apollo 13 was supposed to land on the moon, but moon landing had to be aborted due to an onboard explosion.

Flight	Date(s)	Mission
Apollo 1	January 27, 1967	Tragic loss of three Apollo astronauts
Apollo 4	November 9, 1967	First all-up launch of Saturn V
Apollo 5	January 22, 1968	First test of Lunar Module in space
Apollo 6	April 4, 1968	Final uncrewed Apollo test flight

Apollo flights that flew humans:

The following Apollo flights flew humans during Apollo mission:

Flight	Date(s)	Mission
Apollo 7	October 11-22, 1968	Tested the Command Module
Apollo 8	December 21-27, 1968	First to orbit the moon
Apollo 9	March 3-13, 1969	Tested the Lunar Module
Apollo 10	May 18-26, 1969.	Tested the Lunar Module around the moon
Apollo 11	July 16-24, 1969	First to land on the moon. Two astronauts, Neil Armstrong and Edwin "Buzz" Aldrin took the first steps of any human being on the surface of the moon.
Apollo 12	November 14-24, 1969	Landed on the moon. Alan Bean walked on the surface of the moon and visited Surveyor 3 which had landed nearby two years earlier.
Apollo 13	April 11-17, 1970	Was supposed to land on the moon but moon landing had to be aborted due to an onboard explosion.
Apollo 14	January 31-February 9, 1971	Landed on the moon. Landing in the site originally planned for Apollo 13 Astronauts. Alan Shepard and Edgar Mitchell walked on the surface on the moon, and also played golf there!
Apollo 15	July 26-August 7, 1971	Landed on the moon. Astronauts used the first Lunar Rover during this mission, enabling them to explore a greater area than previous astronauts were able to while on the moon.
Apollo 16	April 16-27, 1972	Landed on the moon. Like Apollo 15, the astronauts used a Lunar Rover to explore the mountainous area of their landing.

Apollo 17	December 7-19, 1972	Landed on the moon. Astronaut Eugene Cernan took the last steps to this date on the moon on December 14, 1972.

Notes:

(1). Apollo 13 was supposed to land on the moon. On the way there, the spacecraft had a problem. NASA had to figure out how to bring the astronauts home safely. Apollo 13 flew all the way around the moon before returning home. Despite the problem, they were able to land safely on earth.

(2). On the last three missions, astronauts drove on the moon with the lunar rover to explore more of the moon's surface. The lunar rovers were made so they could be folded and stored in the Lunar Module. The lunar rovers were left on the moon.

Landing sites for Apollo Moon missions (Apollo 13 was intended to land in the Fra Mauro area, the same area as Apollo 14). Annotations (on the left side) describe the landing sites and the durations on the lunar surface.
Credits: NASA Scientific Visualization Studio
Visualizations by Ernie Wright Released on March 29, 2019.
https://svs.gsfc.nasa.gov/4731

Q218. What spacecraft were used for the Apollo Project?

A218. Rocket and Spacecraft

Rockets: Two types of rockets were used for the Apollo programme. The first flights used the smaller Saturn I B rocket with two stages. When the first part ran out of fuel, it dropped away from the other and burned up in earth's atmosphere. The second part continued flying. The other flights used the more powerful Saturn V (5) rocket. This three-stage rocket sent the Apollo spacecraft to the moon. It was about as tall as a 36-story building.

Apollo Spacecraft: Apollo was a three-part spacecraft:

1. The command module (CM): It was a capsule with room for three astronauts. The astronauts rode in the Command Module on the way to the moon and back.
2. The service module (SM): for the propulsion and spacecraft support systems

   (When CM and SM were together, the two modules are called CSM)
3. The lunar module (LM): to take two of the crew to the lunar surface, support them on the moon, and return them to the CSM in lunar orbit.

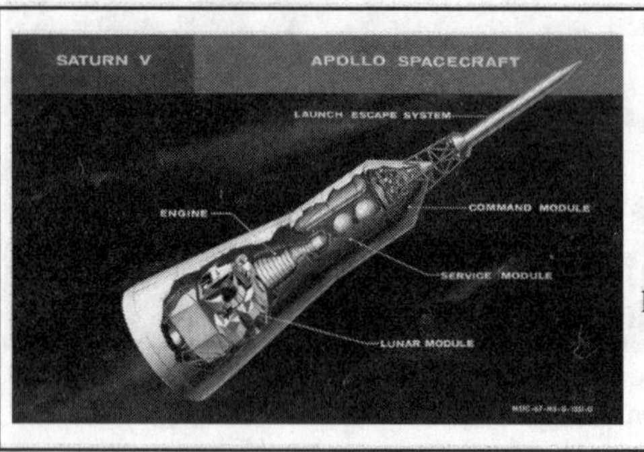

Cutaway view of the Apollo spacecraft configuration aboard the Saturn V rocket. *Credit:* NASA. Date 1967 Source: https://i.imgur.com/6f2EZmL.jpg Wikimedia Commons https://commons.wikimedia.org/wiki/File:Apollo_Spacecraft_diagram.jpg Figure in public domain

Q219. How did Apollo astronauts land on the moon and return to the earth?

A219. The Apollo spacecraft were launched on top of the Saturn V rocket which was made of three stages. The first two stages used up their fuel reaching orbit. The third stage was used to push the Apollo Command Module (CM) and Lunar Module (LM) towards the moon. Once the spacecraft reached the moon's orbit, the two modules separated from each other. Two astronauts in the LM landed on the lunar surface. The third astronaut stayed in the CM in orbit around the moon. When the two astronauts finished working on the moon's surface (exploring the surface, setting up experiments, taking pictures, collecting rock samples, etc.) they got back in the LM, launched it, went back into moon's orbit, and connected with the CM. The two astronauts moved into the CM. They left the LM behind and flew back to earth. The Command Module landed in the ocean, and a ship picked up the astronauts.

Q220. What are some of the milestones and learnings from the Apollo project?

A220. Apollo set several major human spaceflight milestones such as:

1. It stands alone in sending crewed missions beyond low earth orbit.
2. Apollo 8 was the first crewed spacecraft to orbit another celestial body.
3. The programme returned 842 pounds (382 kg) of lunar rocks and soil to earth, greatly contributing to the understanding of the moon's composition and geological history.
4. Apollo also spurred advances in many areas of technology including rocketry, avionics (electronic equipment fitted in an aircraft), telecommunications, and computers. Regarding advances in rocketry, scientists and engineers developed step by step from Little Joe II to a three-stage liquid-fuelled rocket, the Saturn V. As of 2021, the Saturn V remains the tallest, heaviest, and most powerful rocket ever brought to operational status.

Scientists learned a lot from the Apollo project. Some of the learnings are:

1. The moon is an evolved terrestrial object and not primordial. Its internal structure is like that of earth with thick crust and a small iron core. Its crust is thicker on the far side (the side which is not seen from earth). Mass is not distributed uniformly inside the moon. Large mass concentrations ("Mascons") lie beneath the surface of many large lunar basins.
2. The earth and the moon have common elements but in different proportions. The moon is rich in calcium and aluminium and poor in sodium compared to earth.
3. The moon is ancient and still preserves the first billion years of the solar system's history. Moon rock ages range from about 3.2 billion years in the maria to nearly 4.6 billion years in the terrae. The moon has three types of rocks: Basalts (dark lava rocks that fill mare basins), Anorthosites (light rocks that form the ancient highlands), and Breccias (composite rocks formed from all other rock types through crushing, mixing, and sintering during meteorite impacts). The moon does not have sandstones, shales, or limestones.
4. Early in its history, the moon melted to great depths to form a magma ocean. The highlands contain low-density rocks that floated to the surface of the magma ocean. The large, dark basins are gigantic impact craters, formed early in lunar history, that were later filled by lava flows about 3.2-3.9 billion years ago.

Reference

Smithsonian National Air and Space Museum. The Apollo Programme
https://airandspace.si.edu/explore-and-learn/topics/apollo/apollo-programme/apollo-results

Q221. What are NASA's lunar/space exploration activities after the Apollo project?

A221. NASA had following lunar/space exploration projects after the Apollo project.

1. The second series of the Pioneer programme (1965-1992)

The Pioneer programme was a series of United States unmanned space missions that were designed for lunar and planetary exploration. The first series (1958-60) was before the Apollo project and has been discussed earlier.

The second series, ran from 1965 to 1992, sent four spacecraft to measure interplanetary space weather, two to explore Jupiter and Saturn, and two to explore Venus. The two outer planet probes Pioneer 10 and Pioneer 11 became the first of five artificial objects to achieve the escape velocity that will allow them to leave the solar system.

2. Clementine Lunar Probe

Clementine was launched on January 25, 1994. Its objective was to test sensors and spacecraft components in long-term exposure to space and to make scientific observations of both the moon and the near-earth asteroid 1620 Geographos. The lunar observations included obtaining multi-spectral imaging of the entire lunar surface, assessing the surface mineralogy of the moon, obtaining altimetry from 60N to 60S latitude, and obtaining gravity data for the near side. For over two months Clementine mapped the moon, producing the first multispectral global digital map of the moon, the first global topographic map, and contributing several other important scientific discoveries, including the possibility of ice at the lunar South Pole.

Observation of the asteroid was not made due to a malfunction in the spacecraft.

3. Lunar Prospector

Lunar Prospector was designed for a low polar orbit investigation of the moon, including mapping of surface composition and possible polar ice deposits, measurements of magnetic and gravity fields, and study of lunar outgassing events. It was launched on January 6, 1998. It orbited the moon for almost 19 months. The probe found evidence suggesting water ice at both poles. The mission ended with the spacecraft impacting the lunar surface, creating a dust cloud that was studied from earth. Data from the mission allowed construction of a detailed map of the surface composition of the moon, and improved our understanding of the origin, evolution, current state, and resources of the moon.

4. Lunar Precursor Robotic Programme (LPRP)

The LPRP supports America's return to the moon by executing lunar robotic missions to conduct research and prepare for future human exploration. The missions will gather data important for reducing the risks of returning humans to the moon such as examining the lunar radiation environment and identifying terrain hazards (slope, roughness, obstacles). The mission will cover areas of scientific and operational interest. Temperature and lighting conditions over an annual cycle, along with a good characterization of dust, environmental conditions, and radiation are needed for mission and hardware design. Resource identification and mapping will inform decisions about possible future use of in-situ resources.

4.1 Lunar Reconnaissance Orbiter (LRO)

LRO, LPRP's first mission was launched on June 18, 2009. It is a NASA robotic spacecraft currently orbiting the moon in a polar mapping orbit. Its primary goal was to make a 3D map of the moon's surface from lunar polar orbit. LRO will provide critical information about the moon to enable selection of safe landing sites. LRO began its work in lunar orbit in September 2009. Since then, the spacecraft has circled the moon again and again. On each circuit, the scientists watching its data from home can identify differences on the lunar surface as meteorites, and occasionally spacecraft, slam into the moon. The ability to see changes over time is what makes LRO's longevity so valuable to scientists.

4.2 Lunar Crater Observation and Sensing Satellite (LCROSS)

LCROSS was launched with LRO on June 18, 2009. The main objective was to further explore the presence of water in the form of ice in a permanently shadowed crater near a lunar polar region. It was successful in confirming water in the southern lunar crater

Cabeus. It suffered a malfunction on August 22, and on October 9, 2009, the mission's impactor and primary spacecraft impacted the lunar surface near the moon's south pole. The LCROSS data showed that the lunar soil within shadowy craters is rich in useful materials, such as hydrogen gas, ammonia, and methane, which could be used to produce fuel for space missions. Large amounts of light metals, such as sodium, mercury, and silver, were discovered. Hydrogen found at the lunar poles belong to crystals of water ice which concluded that there is water on the moon.

4.3 LADEE

The Lunar Atmosphere and Dust Environment Explorer (LADEE) was a NASA lunar exploration and technology demonstration mission. It was launched on September 7, 2013. During its seven-month mission, LADEE orbited around the moon's equator and studied the lunar exosphere and dust in the moon's vicinity.

The mission ended on April 18, 2014, and intentionally crashed into the far side of the moon.

5. Gravity Recovery and Interior Laboratory (GRAIL)

The GRAIL mission flew twin spacecraft--Ebb and Flow--in tandem around the moon. The launch date was September 10, 2011. The mission was to map lunar gravity and use that information to increase understanding of the moon's interior and thermal history. As the two spacecraft flew over areas of greater and lesser gravity (caused both by mountains and craters and by masses hidden beneath the lunar surface), they moved slightly towards and away from each other. An instrument aboard each spacecraft measured the changes in their relative velocity very precisely and scientists translated this information into a high-resolution map of the moon's gravitational field. Ebb and Flow ended their mission on December 17, 2012, with a controlled crash into the lunar surface. Their final contribution was to create a cloud of dust and gas which researchers could use to discover more about the moon's composition.

Q222. What is NASA's Artemis project?

A222. The Artemis programme is an ongoing international human spaceflight programme led by United States. Its primary goal is of returning humans to the moon by 2025, specifically to the lunar south pole. It is NASA's first crewed moon mission since Apollo 17 in 1972. Long-term objectives are laying the foundations for the extraction of lunar resources, and eventually making crewed missions to Mars and beyond. In collaboration with commercial and international partners, NASA will establish sustainable structures on and around the moon (in lunar orbit) to support more exploration than ever before. The system is being developed in three Blocks: Block 1, Block 1B, and Block 2. The first three Artemis missions will use Block 1, five subsequent flights will use Block 1B, after which all flights will use Block 2.

> Note: Artemis is the mythological Greek goddess of the moon and twin sister of Apollo. The name 'Artemis' is also significant as one of the mission's objectives is to land the first woman on the moon.

Q223. What are the main phases of the Artemis moon project?

A223. There are four main phases of the Artemis moon missions needed to make a crewed moon landing. These include:

1. Orion Spacecraft

The Orion spacecraft is now being designed as the Multi-Purpose Crew Vehicle (MPCV) to launch astronauts into space, perform multiple long distance/long duration space missions, and then return the astronauts safely back to earth. It will provide emergency abort capability. It will have five primary elements:

(a) Crew Module (CM): Houses and transports astronauts during spaceflight missions.

(b) Service Module (SM): Contains Orion's propulsion, power and life support systems.

(c) The Spacecraft Adaptor and Fairings: Connects Orion to the launch vehicle.

(d) Launch Abort System (LAS): Propels the Orion Crew Module to safety in an emergency during launch or ascent.

(e) Multi-Purpose Crew Vehicle to Stage Adaptor (MSA): Connects the entire vehicle structure to the kick stage of the rocket.

Note: Orion is one of the constellations in the sky and the mythological hunting companion of Artemis.

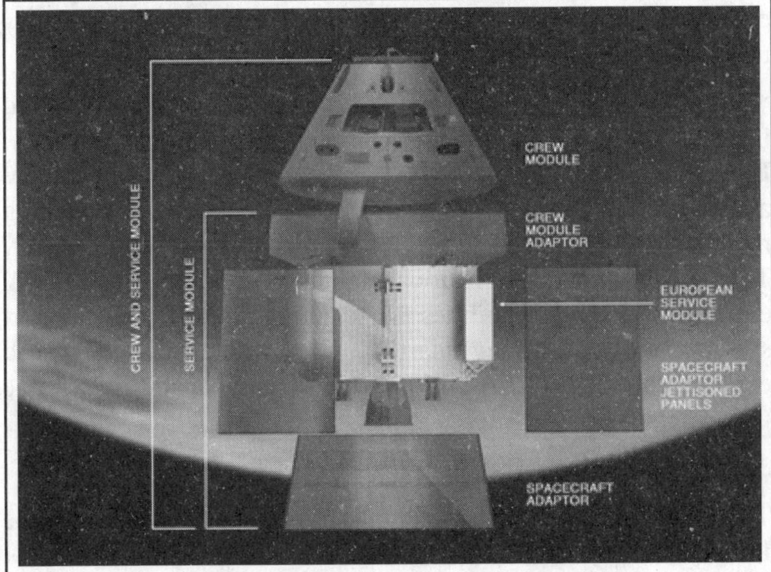

Orion's planned crew, command, and service modules.
https://www.nasa.gov/feature/nasa-gears-up-to-test-orion-s-powerhouse
Credits: NASA (2019)

2. Lunar Gateway

The Lunar Gateway will be a mini-space station in lunar orbit intended to serve as a solar-powered communication hub, science laboratory, short-term habitation module, and holding area for rovers and other robots. It is designed to dock with the Orion module. Launching of the first parts of the structure is planned in 2022. Use of the Gateway in lunar orbit will be one of the key differences between Apollo and Artemis projects. It will be built with commercial and international partners. Operating autonomously when the crew is not present, it also will be a platform for new science and technology demonstrations around the moon. The Gateway is planned to be deployed in a near-rectilinear halo orbit (NRHO) around the moon, a highly inclined orbit around the moon. The NRHO is a seven-day cycle, taking the Gateway as close as approximately 1,600 km (1,000 miles) and as far away as about 68,260 km (42,415 miles) from the lunar surface. Consequently, roughly every seven days, the lunar lander can depart from the Gateway to travel to the lunar surface.

The Gateway is meant to be developed, serviced, and utilized in collaboration with commercial and international partners: European Space Agency (ESA), Canadian Space Agency (CSA), and Japan (JAXA). ESA will contribute habitation and refueling modules and enhanced lunar communications to the Gateway. The ESA-provided International Habitation module, I-HAB, will enhance Gateway capabilities for scientific research, life support systems and crew living quarters. CSA will provide advanced external robotics. JAXA will provide several capabilities for the Gateway's I-HAB.

3. Moon Landing Module

The lunar landing vehicles will take cargo and humans from the Lunar Gateway to the moon's surface. NASA and commercial companies are currently working to develop both a human landing system (known as HLS) and other vehicles. The landing systems for the Artemis missions are set to be used for multiple missions. HLS is a critical component for the Artemis mission. This system transports crew from lunar orbit (the Gateway or an Orion spacecraft) to the lunar surface, acts as a lunar habitat, and then transports the crew back to lunar orbit.

4. Space Launch System (SLS)

Space Launch System (SLS), an advanced super heavy-lift rocket, is the launcher that will carry humans and cargo beyond the earth's atmosphere and into space. This super heavy-lift rocket can carry almost 180,000 kg (397,000 pounds). When complete, the SLS will be the most powerful rocket in the world.

Regarding journey to the moon and beyond, Orion spacecraft, atop the Space Launch System (SLS), will carry four astronauts and will dock with the Lunar Gateway in lunar orbit. The crew, depending on the mission's activity, will have a multi-month stay in the lunar gateway. The human landing system (HLS) will then take two astronauts down to the moon's surface and explore with the habitable mobility platform (a moon base), while the remaining two astronauts stay aboard Gateway. The four crew members are later reunited aboard the gateway for another multi-month stay simulating the return trip to earth.

Q224. How is NASA's Artemis project different from the Apollo project?

A224. We learned a lot from the Apollo missions and that invaluable knowledge is being put to good use in planning a return to the moon with the Artemis project. A few of the key differences between NASA's two lunar programmes, Apollo and Artemis are given below:

1. **Goals:** While the goal of Apollo was to land the first humans on the moon, the Artemis programme will use the moon as a testbed for farther exploration into the solar system, beginning with Mars. In a way, Artemis is America's "moon to Mars" space exploration approach using a multi-month split-crew operation at the Gateway and on the lunar surface. The Artemis mission will have a more continuous and longer presence on the moon. Another key focus will be on developing a lunar economy.
2. **Crew capacity:** While the Apollo command modules carried a crew of three, the Artemis's larger Orion spacecraft will carry four astronauts (Orion will have a capacity of six astronauts).
3. **Technologies:** The technologies involved in the Artemis project is much more advanced than used in Apollo project.

The space launch system (SLS) will generate 15 per cent more thrust than Apollo's Saturn V and will launch Artemis astronauts on missions to the moon, Mars, and beyond. The Orion is designed and built to travel farther than the Apollo capsules. While the moon was the final stop for the Apollo spacecraft, it is the first stop for Orion. Orion's technology is more advanced than the Apollo spacecraft. According to NASA, Orion's computer is 20,000 times faster and its memory capacity is 128,000 times greater.

4. First woman and person of colour on the Moon: The Apollo missions saw no woman or person of colour stepping foot on the moon. With the Artemis flights, NASA has pledged to put the first woman and person of colour on the moon. And It's getting real. On April 3, 2023, NASA announced the four astronauts they will be sending to the Moon as part of its Artemis II mission: three from the US and one from Canada. The team has a woman (Christina Hammock Koch - Mission Specialist) and a person of color (Victor Glover - Pilot). The other two members are: Reid Wiseman (Commander), and Jeremy Hansen (Mission Specialist). They will travel around the Moon and back as part of the 10-day Artemis II mission. NASA is aiming to launch these astronauts in November 2024 at the earliest.

Q225. What are NASA Artemis's planned and future missions?

A225. As of now, Artemis's planned missions are Artėmis I to Artemis III.

Mission	Launch date	Duration	Goal
Artemis I	2022	≈ 25 days	Uncrewed lunar orbit and return
Artemis II	2024	≈ 10 days	4-person lunar flyby

Artemis III	2025	≈30 days	4-person lunar orbit with 2-person lunar landing

Artemis I was an uncrewed Moon-orbiting mission. It was the first integrated flight test of the Orion spacecraft and Space Launch System (SLS) rocket. Its main objective was to test the Orion spacecraft, especially its heat shield, in preparation for subsequent Artemis missions.

Artemis 1 was launched on November 16, 2022. Orion completed one flyby of the Moon on November 21, 2022, entered a distant retrograde orbit for six days, and completed a second flyby of the Moon on December 5, 2022. The Orion spacecraft then returned and reentered the earth's atmosphere with the protection of its heat shield, splashing down in the Pacific Ocean on December 11, 2022. The duration of the mission was 25 days, 10 hours and 53 minutes and distance travelled was 1.3 million miles (2.1 million kilometers)

Artemis II (2024) will be the first crewed test flight of SLS and the Orion capsule. The four crew members will perform extensive testing in earth orbit and Orion will then be boosted into a free-return trajectory around the moon, which will return Orion back to earth for re-entry and splashdown. This will be the first time in 50 years that humans have flown close to the Moon and paves the way for Artemis III when astronauts set foot on the lunar surface.

On April 3, 2023, NASA announced the four astronauts they will be sending to the Moon as part of its Artemis II mission: three from the US and one from Canada.

The team will travel around our Moon and back as part of the 10-day Artemis II mission:

- Reid Wiseman (Commander)
- Victor Glover (Pilot)
- Christina Hammock Koch (Mission Specialist)
- Jeremy Hansen (Mission Specialist)

Artemis II crew; left to right: Christina Hammock Koch, Reid Wiseman (seated), Victor Glover, and Jeremy Hansen. Credit: NASA

Artemis III (2025): Building on the Artemis II mission, four astronauts aboard the Orion module will dock with the Lunar Gateway and remain in space for 30 days. The human landing system (HLS) will then take two astronauts down to the moon's South Pole. The astronauts are expected to spend a week exploring the surface and perform a variety of scientific studies, including sampling water ice. The HLS will then ascend to return the two astronauts to a rendezvous with Orion. Orion will return the four astronauts to earth.

The mission is intended to be the first to place a woman on the moon.

Artemis III has the potential to increase our understanding of how the moon formed and evolved, how it interacts with the sun, and its water and resources. The Artemis III Science Definition Team (SDT) identified following seven objectives:

1. Understanding planetary processes
2. Understanding the character and origin of lunar polar volatiles
3. Interpreting the impact history of the earth-moon system
4. Revealing the record of the ancient sun and our astronomical environment
5. Observing the universe and the local space environment from a unique location
6. Conducting experimental science in the lunar environment
7. Investigating and mitigating exploration risks

Artemis IV, V, VI and more?

NASA is currently focusing attention on Artemis missions I to III. If these prove successful, NASA has ambitions for further crewed missions on an annual basis through to 2030. One expectation is for future astronauts to begin establishing a base on the surface of the moon, with a view to eventually using the moon as a staging post on the journey to Mars.

Q226. What is Artemis accords?

A226. As space becomes easier and easier to reach, more and more countries and private organizations are going to flying to space. We are going to need some rules. There's already the Outer Space Treaty, that's been signed by most nations on earth, but it's mostly concerned with the proliferation of weapons of mass destruction. NASA has developed the Artemis Accords, new rules for space agencies who want to participate in the exploration of space. The Artemis Accords will describe a shared vision for principles, grounded in the Outer Space Treaty of 1967, to create a safe and transparent environment which facilitates exploration, science, and commercial activities for all of humanity to enjoy.

As of March 2023, 24 countries have embraced the Artemis Accords: Australia, Bahrain, Brazil, Canada, Columbia, France, Israel, Italy, Japan, the Republic of Korea, Luxembourg, Mexico, New Zealand, Nigeria, Poland, Romania, Rwanda, Singapore, South Arabia, Ukraine, United Arab Emirates, United Kingdom, Isle of Man, and the United States.

Q227. What is NASA's VIPER lunar rover?

A227. The VIPER (Volatiles Investigating Polar Exploration Rover) is NASA's lunar rover planned to be delivered to the surface of the moon in late 2024. NASA's Artemis programme envisions people sustainably living on the moon using locally available resources like water. We know from past missions that the lunar poles have water ice. But to be able to use that water, we need to learn about its state and how accessible it is. VIPER will fulfil this requirement. It will prospect lunar resources in Permanently Shadowed Areas (PSA) in the lunar south pole region, especially by mapping the distribution and concentration of water ice. It is planned to travel several kilometres, collecting data on different kinds of soil environments affected by light and temperature.

VIPER is a golf cart-sized rover that will be powered by a solar-charged battery. It will roam around the moon's south pole at a top speed of 0.8 km/h and slow down to 0.4 km/h. The rover will be equipped with different drills and science instruments for achieving the goal of hunting water ice. Instruments will look for water below the moon's surface up to 3ft deep and will study the nature of the water. VIPER will also be equipped with a lighting-cum-camera system to allow it to explore inside dark craters. The headlight and camera system will be able to operate in the extreme conditions of light and dark on the moon. Once it enters a permanently shadowed location, it will operate on battery power alone and will not be able to recharge it until it drives to a sunlit area. Its total operation time will be approximately 100 earth days.

VIPER is a planned robotic lunar rover by NASA that will be tasked to prospect for natural lunar resources, especially water ice within a permanently shadowed region near the lunar south pole.
Credits: NASA Ames/Daniel Rutter
Wikimedia Commons
https://commons.wikimedia.org/wiki/File:VIPER_lunar_rover.jpg
Author: NASA. October 23, 2019
Source: https://www.nasa.gov/feature/new-viper-lunar-rover-to-map-water-ice-on-the-moon
Figure in Public domain

Q228. What are NASA's Plans for a Lunar Base Camp?

A228. We might have astronauts living permanently on the moon just like we have astronauts on board the International Space Station (ISS). To meet this requirement, NASA is planning to build a long-term base on the surface of the moon as a part of its Artemis programme. NASA will start constructing it during some of its early missions in the Artemis programme. The following three core elements for a sustained presence have been proposed:

1. Lunar Terrain Vehicle (LTV) that would transport crew around the landing zone.
2. Habitable Mobility Platform (HMP), which would allow

crews to conduct journeys across the lunar surface lasting up to 45 days at a time.

3. Lunar Foundation Surface Habitat (LFSH), which would house as many as four crew members during shorter-duration missions on the surface.

Concept art from NASA showing astronauts entering a lunar outpost. (2006)
Credits: Wikimedia commons.
Author NASA. Date May 15, 2006
https://commons.wikimedia.org/wiki/File:Entering_a_Lunar_Outpost.jpg
Source: http://www.nasa.gov/mission_pages/exploration/multimedia/jfa18833.html
Figure in Public domain

❑

CHAPTER 20

USSR's/ Russia's Lunar Exploration Programme

Q229. What were the Soviet Union's lunar/space exploration missions during 1959-1979?

A229. The Soviet Union (short for the Union of Soviet Socialist Republics or USSR) was the leading country in space exploration in the 1950s and 60s. They had several firsts to their credits. Unlike the USA, they kept their missions secret till they wanted to make an announcement.

In the year 1958, the USSR government approved building a multi-stage rocket required for their moon exploration programme. Their moon missions went under the names Luna and few under Zond. The programme included a series of robotic missions. Soviet engineers developed sophisticated spacecraft capable of bringing moon soil samples to earth and to deploy Lunokhod rover vehicle that explored the moon's surface. However, their human landing mission could not succeed as their super booster, like American Saturn V could not function as expected.

The Luna Missions (1959 – 1976)

A total of 24 missions were designated Luna out of which 15 were successful in accomplishing their missions. These probes were used for the following six types of missions. Successful probes and their launch dates are also mentioned.

1. Impactors: Designed to hit the near side of the moon, transmitting photographs back to earth until their destruction on impact.

Luna 2 (September 12, 1959)

2. Flybys: Designed to transmit photographs back to earth.

Luna 3 (October 4, 1959) rounded the moon and returned the first photographs of its far side, which can never be seen from earth.

3. Soft landers: designed to softly land on the surface of the moon by slowing the speed sufficiently to prevent the craft's destruction. It can continue to transmit pictures from the surface, and possibly dig into the lunar soil or return other information about the lunar environment.

Luna 9 (January 31, 1966) and Luna 13 (December 21, 1966)

4. Orbiters: designed to orbit the Moon

Luna 10 (March 31, 1966), Luna 11 (August 24, 1966), Luna 12 (October 22, 1966), Luna 14 (April 7, 1968), Luna 19 (September 28, 1971), and Luna 22 (May 29, 1974)

5. Rovers: designed to deploy wheeled vehicles to explore a wider area of the lunar surface than the immediate landing site. Luna 17 and Luna 21 carried Lunokhod vehicles. Lunokhod rovers were about 2.3 metres long and 1.5 metres tall.

Lunokhod 1 travelled 10.5 kilometres (6.5 mi) in 322 days and returned more than 20,000 television images and 206 high-resolution panoramas.[5] Lunokhod 2 operated for about four months, covered 42 kilometres (26 mi) of terrain.

Luna 17 (November 10, 1970) and Luna 21 (January 8, 73)

6. Sample return: designed to robotically scoop up a small amount of lunar material, lift off from the surface, and return the material to earth. Three Lunas (16, 20, and 24) returned a total of 301 g (10.6 oz) samples of lunar soil to earth.

Luna 16 (September 12, 1970), Luna 20 (February 14, 1972), and Luna 24 (August 9, 1976)

Model of Luna 24 lunar lander
Credits: Wikimedia Commons
https://commons.wikimedia.org/wiki/File:%D0%9B%D1%83%D0%BD%D0%B0-24_3_(24432633921)_cropped.jpg
Date January 13, 2016,
Author: Музей Космонавтики from Россия (Space Museum from Russia)
Figure in Public Domain

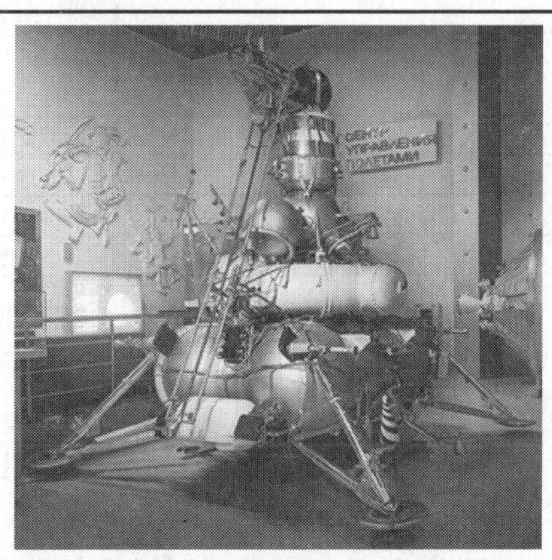

Zond Mission (1964 – 1979)

The Zond Programme was conducted by the Soviet Union to investigate the moon and its vicinity. This programme began in 1964 and ended in 1979. Although most of the Zond flights were oriented towards gathering information required to get astronauts to the moon and back safely, they did collect other information of scientific interest. Instrumentation flown on these missions gathered data on micrometeor flux, solar and cosmic rays, magnetic fields, radio emissions, and solar wind. Biological payloads were also flown, and many photographs were taken. Zond 1 was sent in the direction of Venus and Zond 2 in the direction of Mars. Selected Zond missions are noted in the following list.

The first three missions were intended to explore Venus and Mars. The missions Zond 4 through Zond 8 were test flights for the Soviet's crewed lunar programmes during the moon race. These could carry two cosmonauts. But the Zond spacecraft made only uncrewed automatic flights. Instrumentation flown on these missions gathered data on micrometeor flux, solar and cosmic rays, magnetic fields, radio emissions, and solar wind. Many photographs were taken, and biological payloads were also flown.

Mission	Launch Date	Type of mission	Return to earth date
Zond 3	July 18, 1965	Lunar flyby	continued into heliocentric orbit
Zond 5	September 15, 1968	Circumlunar	September 21, 1968
Zond 6	November 10, 1968	Circumlunar	November 17, 1968
Zond 7	August 7, 1969	Circumlunar	August 14, 1969
Zond 8	October 20, 1970	Circumlunar	October 27, 1970

Zond 5 had a biological payload of two tortoises, wine flies, meal worms, plants, seeds, bacteria, and other living matter was included in the flight.

Zond 9 was planned for July 1969, carrying a crew of two cosmonauts but never flew.

The Soviets eventually abandoned the programme in the wake of the successful Apollo missions. During the 1980s, the Soviets focused their attention on launching spacecraft to Venus and Mars. When the Soviet Union dissolved in 1991, Russia inherited most of its space programme.

Reference

Williams, David R. NASA Space Science Data Coordinated Archive (NSSDCA)/ The Goddard Space Flight Centre (GSFC). Soviet Lunar Missions. Last Updated: September 27, 2005
https://nssdc.gsfc.nasa.gov/planetary/lunar/lunarussr.html

Q230. What were the Soviet Union's "first historical space activities?

A230. Beginning in October 1957 with the launch of the first earth orbiting satellite, USSR scored several historical firsts.

1. First earth-orbiting satellite in history - Sputnik 1 on October 4, 1957.
2. First to carry a living creature into orbit, a dog named Laika - Sputnik 2 on November 3, 1957.
3. First spacecraft to escape earth's gravity. Though it failed to impact the moon as planned but became the first man-made object to go into orbit around the sun – Luna 1 launched on January 2, 1959.
4. First man-made object to reach the moon - Luna 2 launched on September 12, 1959.
5. Took first pictures of the far side of the moon covering 70 per cent of the moon's far side - Luna 3 launched on October 4, 1959.
6. First spaceflight to send animals (two dogs named Belka and Strelka alongside two rats, a rabbit, fruit flies) and a range of plants into orbit and return them safely back to earth – Sputnik 5 on August 19, 1960.
7. First man to reach space - Yuri Gagarin made a single orbit around the earth on 12 April 1961 in Vostok 1and became the. He remained in space for one hour and forty-eight minutes before landing in Saratov Oblast, west Russia.
8. First civilian and first woman in space—Valentina Tereshkova spent almost three days in space orbiting the earth 48 times - Vostok 6 launched on June 16, 1963.
9. First spacewalk - Alexei Leonov left his spacecraft in a specialized spacesuit and conducted a twelve-minute spacewalk - Voskhod 2, the launched March 18, 1965
10. First spacecraft soft landing on the moon - Luna 9 launched on January 31, 1966. An onboard television system began broadcasting and several radio sessions totaled up to about 8 hours of data.
11. First spacecraft into the moon's orbit, 460 orbits in 56 days - Luna 10 launched on March 31, 1966.

12. First spacecraft to circle the moon and return to land on earth. It was also the first to carry earth lifeforms (tortoises, wine flies, meal worms, plants, seeds, bacteria, and other living matter around the moon - Zond 5 launched on September 15, 1968.
13. First probe to bring the moon rock samples to earth, returned 105 grams of moon dust (regolith) to earth - Luna 16 launched on September 12, 1970.
14. First moon rover Lunokhod 1 worked for 10 months covering 10.5 km (6.5 miles) - Luna 17 launched on November 10, 1970. The 1,666-pound rover completed a record-breaking 10.5 kilometres (6.5-mile) trek across the moonscape in 10 months of operation. Despite moving no faster than 1.25 miles per hour, the eight-wheeled robot exceeded the time and distance covered by the Apollo expeditions.
15. First space station - Salute 1 launched on April 19, 1971. Parts of this spacecraft became core segments of the International Space Station (ISS) almost thirty years later in November 2000.

Reference

Zak, Anatoly. 50 Years Later, the Soviet Union's Luna Programme Might Get a Reboot. Popular Mechanics. July 18, 2021. https://www.popularmechanics.com/space/moon-mars/a36984208/soviet-luna-programme-history/

Q231. What are Russia's future lunar exploration projects?

A231. With the collapse of the USSR in 1991, Russia's ambitions for the moon mostly disappeared for three decades. As of 2019, the country's space agency Roscosmos has now announced the following plans for Luna 25 to Luna 30.

Luna 25: Nearly 45 years since the last lunar mission Luna 24, Russian scientists are now planning Luna 25 lander with modern electronics and with a brand-new instrument module with vertically positioned solar panels. It will be going to the moon's south pole, a geological area of great interests. Russian scientists want to study water locked below the surface in permanent ice and to evaluate the dangers posed by sharp fragments of lunar dust. As it lands, the spacecraft will use a European-built camera to advance the European Space Agency's future lunar missions. Luna 25 will perform chemical and spectral analysis of the surrounding regolith with the primary goal of finding lunar water.

The country's space agency Roscosmos has announced that Luna 25, which was scheduled to launch in July 2022, is now being ready for a "preferred" July 23 departure.

Luna 26: Russia plans to launch Luna 26 two years after Luna 25. It will be an orbiter that would look for magnetic and gravitational anomalies in the moon and capture high-precision images of potential landing sites.

Luna 27: It would be back to the surface of the moon one year after Luna 26 with collaboration by the European Space Agency (ESA). Luna 27, a lander, will target the moon's South Pole–Aitken basin. The purpose is to prospect for minerals, volatiles (nitrogen, water, carbon dioxide, ammonia, hydrogen, methane and sulfur dioxide, and lunar water ice) in permanently shadowed areas of the moon and investigate the potential use of these natural lunar resource.

The mission is a continuation of the Luna-Glob programme. In addition, the lander will study how the solar wind affects the lunar surface.

Note: UKRAIN WAR: In March/April 2022, The European Space Agency (ESA) said that the ESA will "discontinue cooperative activities" with the Russian space agency Roscosmos on the Luna 25, 26 and 27 moon exploration missions citing Russian aggression against Ukraine and sanctions placed on Russia in its reasoning. ESA and Roscosmos were planning to launch the three missions starting later in 2023 with Luna 25. Luna 26 was scheduled to follow two years later and Luna 27 the following year.

Luna 28: Luna 28, also known as Luna-Grunt rover, is a proposed sample-return mission from the south polar region of the moon. It would be composed of a stationary lunar lander and a small lunar rover. It is planned to be launched in 2025–2027. It would be bringing back to earth cryogenically stored samples from the lunar south pole. The samples will be not just regolith [lunar dirt] but all volatiles and cryogenic inclusions to it. The orbital module will continue its mission in the lunar orbit for at least three years.

Luna 29: It is planned to be launched in 2028. It would carry a new Lunokhod rover and will be a sample return mission.

Note: Russia also has plans for Luna 30 and 31 by 2030.

Q232. What are Russia's and China's joint space /lunar exploration projects?

A232. Russia and China have announced the intent to develop an International Lunar Research Station (ILRS) by 2035 at the south pole of the moon. The ILRS will include an earth-to-Lunar flight system, an auxiliary long term operation system on the lunar surface. The programme will also include plans to support the two countries' national satellite systems.

The two countries are planning to join forces to achieve new heights in space exploration. This could be a sign of a two-track approach to lunar exploration: one led by the United States, along with Canada, Europe, and Japan, establishing a "Gateway" space station around the moon (in lunar orbit) and the other led by China and Russia, ILRS. Some believe that this may be a start of a new "space race"

Russia has also hinted that they may also participate in China's "Tiangong space station" planned to be built in low earth orbit.

China is also developing a list of mutually beneficial projects with Russia that will be included in the new Cooperation Programme for 2023-2027.

On March 9, 2021, the two countries have signed agreements to jointly create a data center for the exploration of the Moon and deep space. They plan to cooperate in the future on China's Chang'e-7 and Russia's Luna 27 missions, which both aim to survey the lunar south pole.

On November 29, 2021, the Russian Chinese "Roadmap for Cooperation in Satellite Navigation for 2021-2025" was signed, which includes plans for the integrated and

Innovative development of Russian GLONASS and Chinese BriDou (BDS) navigation systems.

Notes:

1. The Gateway, a vital component of NASA's Artemis programme, will serve as a multi-purpose outpost orbiting the moon that provides essential support for long-term human return to the lunar surface and serves as a staging point for deep space exploration.
2. GLONASS is a Russian space-based satellite navigation system. It provides an alternative to GPS and is the second navigational system in operation with global coverage and of comparable precision.
3. The BeiDou Navigation Satellite System is a Chinese satellite navigation system consisting of two separate satellite constellations

References

1. *Foust, Jeff. Russia continues discussions with China on lunar exploration cooperation. Space News. April 4, 2021* https://spacenews.com/russia-continues-discussions-with-china-on-lunar-exploration-cooperation/
2. *Xiaoci, Deng and Anqi, Fan. Exclusive: China, Russia to sign new 5-year space cooperation programme, build intl lunar station by 2035: Roscosmos. Global Times. Dec 29, 2021.* https://www.globaltimes.cn/page/202112/1243731.shtml

Moon Landing Map showing landing sites of Surveyor, Apollo and Luna missions
Credits: Wikimedia Commons, NASA
https://commons.wikimedia.org/wiki/File:moon_landing_map.jpg
source: NASA
http://nssdc.gsfc.nasa.gov/planetary/lunar/moon_landing_map.jpg
Figure in public domain

❑

CHAPTER 21

China's Space and Lunar Exploration Programme

Q233. When did China start its space programme?

A233. The space programme of China is directed by the China National Space Administration (CNSA). Its major initial highlights are:

1. China began a ballistic missile programme in the 1950s, first efforts at human spaceflight started in the 1960s, and successfully launched an artificial satellite in April 1970.
2. The first uncrewed spacecraft flight, first unmanned Shenzhou 1, was launched on November 19, 1999.
3. China's first crewed launch took place on October 15, 2003, when Yang Liwei became the first person sent into space. This mission, Shenzhou 5, made China the third country to independently send humans into space. Yang Liwei orbited the earth 14 times, all within 21 hours, 22 minutes, and 45 seconds.

Q234. What is China's Shenzhou programme?

A234. The Shenzhou (means divine vessel) programme is China's Manned Space Programme (CMS) which has been in operation since 1992. The initial plan had three phases:

Phase 1: To launch and return two uncrewed spacecraft, followed by the first Chinese crewed spaceflight by 2002.

Phase 2: To prove the technology, conduct rendezvous, and docking operations in orbit; and operate a spacelab by 2007.

Phase 3: To orbit a long-term space station in the 2010–2015 period.

China signed a deal was with Russia in March 1995 to transfer technology including training of cosmonauts.

As of September 2021, China has successfully completed the first two phases, though there had been delays from what was initially planned. This made China the third nation to master human spaceflight after the Soviet Union/Russia and the United States. The third phase (the construction of Tiangong space station) is ongoing. Ever since the first spacecraft was launched in 1999, CMS has successfully completed 18 missions. Some of the details are:

The first four uncrewed test flights happened between 1999 and 2002. These were followed by crewed missions. The fifth launch, Shenzhou 5, was the first to carry a human, Yang Liwei which occurred on October 15, 2003. It took 14 earth orbits. In the year 2008, a three-person crew was sent to space conducting the first spacewalk.

In 2012, first Chinese woman was sent to space.

Between 2011 and 2021, China has sent several spacecraft for constructing its space station, Tiangong, now scheduled to be completed in 2022-23.

Q235. What were China's space laboratories Tiangong-1 and Tiangong-2?

A235. Tiangong-1 and Tiangong-2 were China's two space laboratories for conducting test experiments towards a bigger plan of making Tiangong Space Station. The first space laboratory, Tiangong-1 was launched in September 2011, then a more advanced space laboratory, complete with cargo spacecraft, Tiangong-2, in September 2016.

Tiangong-1 was China's first prototype space station. It was the first operational component of the "Tiangong Space Station programme". It orbited earth from September 2011 to April 2018. It demonstrated orbital rendezvous and docking capabilities during its two years of active operational life. The uncrewed Shenzhou 8, successfully docked with the module in November 2011, while the crewed Shenzhou 9 mission docked in June 2012. A third and final mission to Tiangong-1was the crewed Shenzhou 10 which docked in June 2013. Tiangong-1 met a fiery end in earth's atmosphere on April 1, 2018, breaking apart and burning up in the skies over the southern Pacific Ocean.

Tiangong-2 was launched on September 15, 2016, and deorbited on July 19, 2019. It was intended as a testbed for key technologies used in the Tiangong space station. Aboard the Shenzhou 11, two astronauts formed the inaugural crew for the Tiangong-2. During the 30 days, the two astronauts conducted a number of scientific and technical experiments on the physiological effects of weightlessness, tests on human-machine collaboration on in-orbit maintenance technology and released an accompanying satellite successfully.

On April 22, 2017, the cargo vessel Tianzhou-1 successfully docked with Tiangong-2 marking the first successful docking and refueling with the orbiting space laboratory. It subsequently performed a second docking and refueling on 15 June 2017. On September 12, 2017, Tianzhou-1 performed the third and final docking and refueling with Tiangong-2, with what is termed a fast docking which took 6.5 hours, rather than 2 days, to complete.

The station subsequently made a controlled reentry on July 19, 2019 and burned up over the South Pacific Ocean.

Q236. What is China's Tiangong space station?

A236. Chinese Manned Space Agency (CMSA) is building an orbiting space station, Tiangong (means Palace in the Sky), in low earth orbit.

It is said that China's incentive to build its own space station was amplified after China was barred from participation in the International Space Station (ISS). It is part of an overall drive to put China on track for more ambitious missions and provide opportunities for cooperation with Russia and other European countries along with the United Nations Office for Outer Space Affairs.

China launched Tianhe (means Harmony of the Heavens), the first of the orbiting space station's three modules in May 2021 and the country aims to finish building the station by end 2022. Once the construction is completed, Tiangong Space Station will enter the application and development phase in 2023, which is poised to last for no less than 10 years and could even be extended to 15 years. Tiangong will continuously house three astronauts for at least a decade.

Being China's first long-term space station, it is the goal of the "Phase 3" of the China Manned Space Programme.

Q237. What are some of the salient features of China's space station?

A237. 1. China's space station, named Tiangong (means Heavenly Palace), is being constructed by China in low earth orbit between 340 and 450 km (210 and 280 mi) above the surface and plans to complete it by the end of 2022.

2. China is building its station independently, though it has invited other countries to contribute experiments and may, in the future, welcome foreign astronauts.
3. The Chinese space station is designed to be used for at least a decade, although the mission could be extended.
4. Once completed, the Tiangong Space Station will have a mass between 80 and 100 tons (180,000 and 220,000 lb),

roughly one-fifth the mass of the International Space Station. The length will be about 20 metre and diameter 4.2 metre.

5. Tiangong will consist of three modules (compared with 16 modules on the International Space Station, ISS):

(a) Tianhe (means "Harmony of the Heavens") - the main habitat for astronauts, and two modules Mengtian (means Dreaming of the Heavens) and Wentian (means 'Quest for the Heavens) dedicated to hosting experiments. China launched Tianhe in April 2021. The other two modules are due to be launched in 2022. The space station could potentially be expanded to six modules.

6. Tiangong will be inhabited by three astronauts. The space station will host many experiments from both China and other countries.

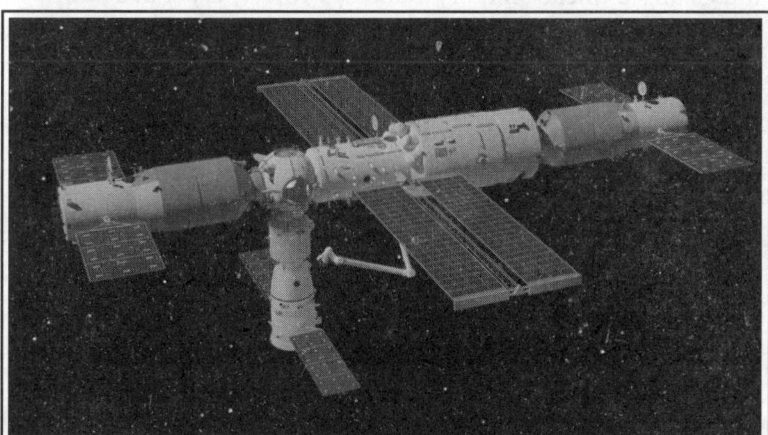

Tiangong Space Station in October 2021, with Tianhe core module in the middle, Tianzhou-2 cargo spacecraft on the left, Tianzhou-3 cargo spacecraft on the right, and Shenzhou-13 crewed spacecraft at nadir.
Credits: Wikimedia Commons.
Author: Shujianyang. October 15, 2021
https://commons.wikimedia.org/wiki/File:Tiangong_Space_Station_Rendering_2021.10.png
Figure is free to share – to copy, distribute and transmit the work

7. Tiangong will have international participation, with China, Russia and Europe intending to keep a cooperative and

multilateral approach in space. Tiangong also involves cooperation from France, and Sweden. European Space Agency (ESA) started human spaceflight training with the ultimate goal of sending ESA astronauts onto Chinese space station. Researchers around the world are eagerly awaiting the completion of Tiangong, which is preparing to host about 1,000 scientific experiments. The space station is expected to host experiments from many countries like Belgium, France, Germany, India, Italy, Japan, Kenya, the Netherlands, Mexico, Peru, Russia, Saudi Arabia, Spain, Switzerland, Poland, Germany, and Norway.

Q238. What are Chinese lunar exploration activities?

A238. The Chinese Lunar Exploration Programme, known as the Chang'e (means moon Goddess) Project, is carried out by the China National Space Administration (CNSA) with the ultimate objective of a crewed lunar landing in the 2030s and possibly build an outpost near the lunar south pole. The project incorporates lunar orbiters, landers, rovers, and sample return spacecraft.

The project is divided into four phases:

Phase I: Orbital missions (Chang'e 1 and Chang'e 2)

In the first phase two lunar orbiters, Chang'e1 and Chang'e 2 were launched in October 2007 and October 2010 respectively.

Chang'e 1 was planned for 1 year and achieved 1 year, 4 months. Chang'e 1 scanned the entire moon and generated a high-definition 3D map. It also mapped the distribution of various chemical elements on the lunar surface. Finally, on March 1, 2009, it was commanded to impact the lunar surface,

Chang'e 2, mapped the moon in even greater detail and to test out technology needed for soft landing, the probe left lunar orbit for the earth–sun L2 Lagrangian point (1.5 million kilometres away from the earth's surface) to test the Chinese tracking and control network. It started orbiting L2 in August 2011. At L2, it studied earth's magnetic field and the solar wind. After orbiting L2 for about 8 months, and transferring data to China National Space Administration, Chang'e 2 departed L2 in April 2012 to begin an extended mission to the asteroid 4179 Toutatis. As of 2014, Chang'e 2 has travelled over 100 million km from earth. Contact with the spacecraft was lost in 2014 as its signal strength weakened due to distance. The probe is expected to return to earth's vicinity sometime around 2027.

Notes:

1. Lagrangian points are locations in space where gravitational forces and the orbital motion of a body balance each other. These were discovered in 1772 by the French mathematician Louis Lagrange. There are five Lagrangian points in the sun-earth system, and Chang'e-2 flew to one nearest earth.
2. 4179 Toutatis is an elongated, stony asteroid and slow rotator approximately 2.5 kilometres in diameter.

Phase II: Soft landers/rovers (Chang'e 3 and Chang'e 4)

Chang'e 3 was launched on 2 December 2013 which landed on the moon on 14 December 2013. It carried a six wheeled lunar rover named Yutu (means Jade Rabbit, the pet rabbit of the Chinese moon goddess Chang'e). The rover was designed to explore an area of 3 square kilometres (1.2 square miles) with a maximum travelling distance of 10 km (6.2 mi) during a 3-month mission. The rover took direct measurements of the lunar soil down to a depth of 30 m (98 ft) and investigated the lunar crust structure. The rover suffered from a mechanical abnormality and ceased to transmit data in March 2015. However, the lander and its telescope are still operational as of September 2020. The power source for the lander could last for 30 years.

Chang'e 3 discovered a new type of basaltic rock, rich in ilmenite, a black mineral.

Chang'e 4, with Yutu-2 rover, was the first mission to land on the far side of the moon. It was launched on December 7, 2018 and landed on January 3, 2019 at Von Karman crater in the South Pole-Aitken Basin.

The rover discovered an unusually coloured, 'gel-like' substance which could be melted glass created from meteorites striking the surface of the moon.

The rover discovered minerals that appeared different from typical lunar surface material. These could be excavated from below the South Pole-Aitken Basin floor by the impact. Analysis revealed the presence of olivine and low-calcium pyroxene and may support the prevailing model of lunar formation and evolution with a cooling magma ocean.

There's Now a fully operational radio telescope on the far side of the moon. The Netherlands-China Low-Frequency Explorer is mounted on the Chinese-built Queqiao communications relay satellite, which relays data from the Chang'e-4 lander and Yutu-2 rover.

Chang'e 4 sees the first cotton seed sprout on the Moon. Cotton seeds carried by China's Chang'e-4 lunar lander have germinated inside the probe becoming the first for any biological matter to grow on the moon. The Chang,e 4 lander carried the seeds of cotton, oilseed rape, potato and arabidopsis, as well as eggs of the fruit fly and some yeast, to form a simple mini biosphere. Cotton was the only seed to germinate so far. However, they died due to the bitter cold of the lengthy lunar night. The canister apparently does not have a battery-powered heater.

The Chang'e-4 mission is still working.

Notes:

1. The South Pole-Aitken Basin is a 2,500-kilometre (1,550 mile) wide, 12-kilometre (7.5 miles) deep ancient impact crater of intense scientific interest which could contain exposed material from the moon's upper mantle and clues to the history and development of the moon. Von Karman is a large lunar impact crater, about 180 km in diameter and lies within the South Pole–Aitken basin.
2. Olivine and Pyroxene are silicate minerals, a large group of silicon-oxygen compounds that are widely distributed

throughout much of the solar system. The silicates make up about 95 per cent of earth's crust and upper mantle. They also are important constituents of lunar samples, meteorites, and most asteroids. In addition, they have been detected on the surfaces of Mercury, Venus, and Mars. Out of several hundred varieties, Olivine and Pyroxene are some of the most significant in rock formation.

Olivine is the name of a group of silicate rock-forming minerals. Besides Silica, they usually have magnesium and iron. They are usually green in colour. During the slow cooling of magma crystals of olivine may form and then settle to the bottom of the magma chamber because of their relatively high density.

Pyroxene is a large group of inosilicates (chain silicate) minerals. They are the most significant and abundant group of rock-forming minerals. Low-calcium pyroxene (often abbreviated as LCP) is an important term for pyroxene low in calcium and high in magnesium and iron.

The Chang'e 4 lander and the ramp designed for Yutu-2 rover deployment.
Credits: Wikimedia Commons. Source: ChangE-4 – PCAM
Author: CSNA/Siyu Zhang/Kevin M. Gill
31 May 2021 (Uploaded)
Free to share – to copy, distribute and transmit the work
https://commons.wikimedia.org/wiki/File:ChangE-4_-_PCAM_(51216550288).png

Phase III: Sample-Return

Chang'e 5 was China's first lunar sample-return mission. It was launched on November 23, 2020, landed on the moon on December 1, 2020, and returned to earth with 1.73 kilograms (3.81 pounds) of lunar soil on December 16, 2020.

Just hours after landing, the probe started using its robotic scoop and drill to dig up material at the landing site, Mons Rumker, a lava dome in a region called Oceanus Procellarum (or the Ocean of Storms) in the northwest lunar near side. This area is characterized by some of the youngest mare basalts on the moon (~1.21 billion years old) with elevated titanium, thorium, and olivine abundances which have never been sampled by Apollo or Luna mission. This data could give new information about the geological evolution of the earth-moon system.

Scientists have begun studying the samples. Some of the early findings are:

1. Scientists have determined the age of basalt fragments to be around 1.96 billion years. That confirms the moon was volcanically active at the time. It makes Chang'e 5 samples the youngest ones collected to date.
2. The Chang'e-5 samples are very diverse, and includes some sharp, jagged lunar particles, silicas, salts, volcanic glasses, and impact glasses, along with different minerals and different rock types. These fragments could give insights into past episodes of energetic, fountain-like volcanic activity on the moon. some materials could have been ejected from as far as 1,300 km away from the Chang'e-5 landing site.
3. Chang'e-5 lunar lander has found evidence of hydroxyl (OH) on the moon.
4. The samples have a lower abundance of heat-producing elements such as uranium, potassium, thorium, etc. than scientists expected. This means further studies are required to understand the history of volcanic activities.
5. Scientists have found water trapped in glass beads. Most of these beads formed over 3 billion years ago. Scientists

think said water is primarily created due to the solar wind interacting with the Moon's surface and is then diffused into the glass beads. Scientists estimate 270 trillion kilograms of water exist in the top 12 meters across the Moon. Just for contrast, Lake Superior—Earth's largest freshwater lake—alone hosts over 11,000 trillion kilograms of water.

References

1. *Atkinson, Nancy. Chang'e-5 Returned an Exotic Collection of Moon Rocks. Universe Today. September 26, 2021.* https://www.universetoday.com/152718/change-5-returned-an-exotic-collection-of-moon-rocks/
2. *Mehta, Jatan. Monday Issue #48. October 11, 2021.* https://blog.jatan.space/p/moon-monday-issue-48

Phase IV: Lunar Research Station

This phase is for the development of a robotic research station near the moon's south pole.

Chang'e 6 - Sample Return (Launch: 2025)

The Chang'e 6 mission is a planned lander designed to return samples from the lunar south pole. It was designed as a backup for Chang'e 5. It aims to bring about two kilograms of lunar samples to Earth. The mission's landing site will be within the farside South Pole-Aitken basin, the largest and deepest lunar crater. It will investigate the topography, composition, and subsurface structure of the landing site.

Chang'e 7 - Orbiter, Lander, Rover, Probe (Launch: 2026)

Chang'e 7 will consist of a lander that will explore the south pole for resources. The mission will include an orbiter, a lander, a rover, and a mini-flying probe. It is planned to make detailed surveys of the south polar region of the moon, covering the terrain, geological composition, locations of water ice, and space environment. One of the primary objectives will be to detect water ice in permanently shadowed areas and determine its origin. The lander will also carry UAE's second rover, which will sport high-resolution cameras from the French space agency. The China National Space Agency will finalize more international payloads to be onboard Chang'e 7. The orbiter and some of the upcoming Russian Luna missions will help select a suitable site for the joint Sino-Russian long-term scientific base on the Moon's south pole, called the International Lunar Research Station (ILRS).

Chang'e 8 - Technology Tests (expected to launch in 2028)

Chang'e 8 is designed to test technologies necessary to the construction of a lunar science base. It will also conduct surveys and scientific experiments. It will verify the utilization and development of natural resources. It may include a lander, a rover, and a flying detector, as well as a 3D-printing experiment using in-situ resource utilization (ISRU) to test-build a structure. It will also transport a small, sealed ecosystem experiment. It will test technology necessary to the construction of a lunar science base.

China is planning to take two astronauts to the lunar surface by around 2030.

Reference

Williams, Dr. David R. Future Chinese Lunar Missions. NASA Goddard Space Flight Center. Last Updated: September 3, 2021. https://nssdc.gsfc.nasa.gov/planetary/lunar/cnsa_moon_future.html

Q239. What is Chinese next-generation reusable crewed spacecraft?

A239. The next-generation crewed spacecraft is a type of reusable spacecraft developed and manufactured by China Aerospace Science and Technology Corporation (CASC). The prototype underwent its first uncrewed test flight on May 5, 2020.

The crew carrier is designed to ferry astronauts to the Chinese space station in earth orbit as well as conducting lunar exploration.

It is larger than the Shenzhou spacecraft that is currently used. The new capsule extends its spacecraft's reliability and safety qualities, and it is reusable.

China's moon Exploration Capsule consists of two modules:

1. A crew module that returns to earth, and
2. An expendable service module to provide propulsion, power and life support for the crew module while in space.

It can carry six astronauts, or three astronauts and 500 kg of cargo. The new crew module is partially reusable with its detachable heat shields, while the spacecraft features a modular design that allows it to be constructed to meet different mission demands.

According to Chinese space officials, the new capsule extends its spacecraft's reliability and safety qualities, and it is reusable. It is China's largest return and re-entry spacecraft with the largest propellant with a launch mass of 21.6 tons.

❑

CHAPTER 22

India's Space and Lunar Exploration Programme

Q240. When did India start space exploration programme and what are its major activities?

A240. Indian Space Research Organization (ISRO) is the space exploration agency of the Government of India. The prime objective of ISRO is to develop space technology and its application to various national needs.

Historically, space research in India began in the 1920s with the studies conducted by the scientists S. K. Mitra, C. V. Raman, and Meghnad Saha. Space related activities were initiated during the early 1960s.

1963 marks the beginning of the Indian space programme when the first rocket was launched for probing upper atmospheric regions. The Indian Space Research Organization (ISRO) was formed in 1969 with an objective to develop space technology and its application to various national needs. It is one of the 6 largest space agencies in the world today.

Satellites

On April 19, 1975, Aryabhata, the first Indian Satellite, completely designed in the country, was launched into space from a Russian facility.

Bhaskara-I, the first experimental remote-sensing satellite, built in India was launched in 1979.The images send by it are used to study hydrology, forestry, and oceanographic studies.

During the 1980s various communication and remote sensing satellites like Rohini, Insat-1A, SROSS-1, and Satellite IRA-1A were launched.

In 1984, a joint manned mission of India and the Soviet Union was launched in which the first Indian cosmonaut, Rakesh Sharma, spent eight days in Russian space station Salyut 7.

Some of the important satellites launched by ISRO are:

Kalpana-1, a meteorological satellite was launched on 12 September 2002.

GSAT-2 was an experimental communication satellite

GSAT-3, also known as EDUSAT, was a communications satellite which was launched on September 20, 2004. EDUSAT is the first Indian satellite built exclusively to serve the educational sector.

Cartosat-1 is a stereoscopic earth observation satellite. The first of the Cartosat series of satellites was launched on May 5, 2005. So far (till November 2019), the ISRO has orbited eight Cartosats. Data from most of them, are exclusively used by the armed forces.

Rockets

During the late 1980s and 1990s, ISRO developed different levels of rockets, the launch vehicles used in several launches.

ASLV, the Augmented Satellite Launch Vehicle or Advanced Satellite Launch Vehicle, a five-stage solid-fuel rocket Small-lift launch vehicle,

PSLV, the Polar Satellite Launch Vehicle, an expendable medium-lift launch vehicle,

GSLV, Geosynchronous Satellite Launch Vehicle, is an expendable launch system. It is capable of placing 2 ton class communication satellite into Geosynchronous Transfer Orbit (GTO). India is one among six countries in the world to demonstrate such launch capability to GTO with the use of complex cryogenic technology. GSLV was used in fourteen launches from 2001 to 2021, with more launches planned. Even though GSLV Mark III shares the GSLV name, it is an entirely different launch vehicle. It is also referred to as the Launch Vehicle Mark 3. It is a three-stage medium-lift launch vehicle.

Launches of Chandrayaan 1 and 2, and Mangalyaan

Chandrayaan-1 - the first moon mission - launched on October 22, 2008

Mangalyaan - India's first interplanetary mission to the planet Mars known as Mars Orbiter mission (MOM) or Mangalyaan – launched on September 24, 2014

Chandrayaan-2 - the second lunar exploration mission - launched on July 22, 2019. The lander of Chandrayaan-2 crashed (caused by a software glitch) when it deviated from its intended trajectory while attempting to land on September 6, 2019. The orbiter is currently in a 100 km × 100 km orbit around the Moon. As of September 2021, it has completed 9,000 revolutions of the moon.

> Note: You may like to refer Shikha Goyal's article in Jagaran Josh (July 15, 2019) giving a timeline of ISRO's space programme. The link is given below.

Reference

Goyal, Shikha. ISRO's Space Programme: Timeline from 1960s to 2019. Jagaran Josh. July 15, 2019. https://www.jagranjosh.com/general-knowledge/isros-space-programme-timeline-1562938305-1

Q241. What is India's Chandrayaan (lunar exploration) programme?

A241. India's lunar exploration programme, also known as Chandrayaan programme, is an ongoing series of space missions by the Indian Space Research Organization (ISRO). It has following three phases:

Phase I: Orbiter and Impactor: Chandrayaan-1

Chandrayaan-1 was launched on October 22, 2008. The mission was a big success as the moon Impact Probe (MEP) discovered water on the moon. The mission also mapped the moon and did moon's atmospheric profiling.

Phase II: Soft landers and rovers

Chandrayaan-2 was conceived with Russian collaboration. However, when Russia expressed its inability to provide the lander, India decided to develop its own and launched Chandrayaan-2 on July 22, 2019. The spacecraft was put into lunar orbit on August 20, 2019. But the lander was lost while attempting to land on September 6, 2019. The orbiter is operational, collecting scientific data and is expected to function for 7.5 years.

Chandrayaan-3 would be a re-attempt to demonstrate the landing capabilities. This spacecraft would have a lander and a rover. The third mission would land in the same area as Chandrayaan 2 was planned to land. As of February 2023, Chandrayaan-3 is scheduled to launch in later part of 2023.

Phase III: Lunar Polar Exploration Mission (LUPEX) - On site sampling

The next mission will be a Lunar Polar Exploration Mission, in collaboration with Japan. It is suggested to be launched in 2024. A Possibility of NASA's participation is also discussed. Details are still being worked out. Most likely it will be a lander-rover mission. It will explore the south pole region of the moon and perform on site sampling and analysis of collected samples and demonstrate lunar night survival technologies.

Reference

From Wikipedia, the free encyclopedia

https://en.wikipedia.org/wiki/Chandrayaan_programme

Q242. What are some of the salient features of Chandrayaan-1?

A242. Chandrayaan-1 was the first Indian planetary science and exploration mission of India. It was launched on October 22, 2008 and was operational for 312 days till August 28, 2009. The spacecraft made more than 3400 orbits around the moon and the mission was concluded when the communication with the spacecraft was lost. This was short of the spacecraft's planned two-year lifetime. ISRO noted that at least 95 per cent of its mission objectives had been accomplished by then

The mission was aimed at high-resolution Remote Sensing of the Lunar surface with following specific scientific goals:

1. To prepare a three-dimensional atlas of both near and far side of the moon.
2. To conduct chemical and mineralogical mapping of the entire lunar surface which will test the hypothesis for the origin and early evolutionary history of the moon and help in determining the nature of the lunar crust.

The spacecraft carried 11 scientific instruments. Five of them were Indian and other six were from: ESA (UK, Germany, Sweden) -3, NASA -2 and Bulgarian Academy of Sciences - 1.

It carried NASA's imaging spectrometer, moon Minerology Mapper (M3), which helped confirming the discovery of water locked in minerals on the moon. The orbiter also released an impactor that deliberately crashed into the moon, releasing debris that was analysed by the orbiting spacecraft's science instruments.

Chandrayaan-1 played a crucial role in the discovery of water molecules on the moon. It detected water (H2O) and hydroxyl (OH) on the lunar surface. The data also revealed their enhanced abundance towards the polar region.

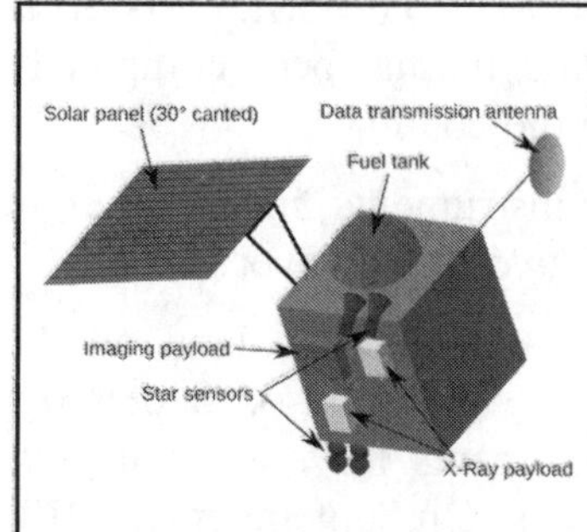

Diagram of the Chandrayaan-1 lunar orbiter
Credits: Wikimedia Commons
https://commons.wikimedia.org/wiki/File:Chandrayaan-1.svg
Author: Reproduction by User: Waterced of an image from NASA.
June 23, 2018
Source: https://nssdc.gsfc.nasa.gov/nmc/spacecraftDisplay.do?id=2008-052A
Figure in Public Domain

References

1. Department of Space, Indian Space Research Organization (ISRO). CHANDRAYAAN – 1 India's First Lunar Exploration Mission
https://www.isro.gov.in/pslv-c11-chandrayaan-1
2. *NASA Science – Solar System Exploration. Chandrayaan-1 /Moon Impact Probe. Page Updated: April 5, 2019.*
https://solarsystem.nasa.gov/missions/chandrayaan-1/in-depth/

Q243. What are some of the salient features of Chandrayaan-2?

A243. Chandrayaan-2 is India's second lunar exploration mission after Chandrayaan-1. It had a lunar orbiter, a lander (named Vikram), and a six wheeled rover (named Prgyan), all of which were developed in India. The primary objectives of the Chandrayaan-2 lander were to demonstrate the ability to soft-land and operate a robotic rover on the lunar surface. The scientific goals of the orbiter are to study lunar topography, mineralogy, elemental abundance, the lunar exosphere, and signatures of hydroxyl and water ice, to study the water ice in the south polar region and thickness of the lunar regolith on the surface, and to map the lunar surface and help to prepare 3D maps of it

The Chandrayaan-2 was successfully launched on July 22, 2019 and inserted into an elliptical orbit around the moon on August 20, 2019. It was carried by the Geosynchronous Satellite

Launch Vehicle-MkIII (GSLV Mk-III M1). The GSLV Mk-III is India's most powerful launcher to date and has been completely designed and fabricated from within the country.

The satellite carried 14 scientific instruments, including eight in the orbiter, four in the lander and two in the rover.

The lander and the rover were scheduled to land on the near side of the moon, in the south polar region on September 6, 2019 and conduct scientific experiments for one lunar day, which approximates to two earth weeks. On September 2, 2019, the Vikram lander separated from the Orbiter and landing was attempted on 7th September. However, the lander deviated from its intended trajectory and missed the primary scheduled landing position. The Vikram lander went for a second attempt, and that is when the visuals went missing. According to ISRO, communication from the lander Vikram was lost. As per a failure analysis report submitted to ISRO, a software glitch seems to have caused the crash. However, the Orbiter is healthy, and all the payloads are operational.

If the lander had landed successfully at the precisely targeted spot, the rover would have rolled on the lunar surface and collected data and images to send back to the Erath for analysis. In its fourteen-earth days (which is one lunar day) lifespan, it could have travelled 500m.

The orbiter part of the mission, with eight scientific instruments, remains operational, and will continue its seven-year mission to study the moon. Based on published data, it can be estimated that as of April 2023 Chandrayaan-2 lunar orbiter has completed more than 16000 trips around the Moon at a rate of about 14 orbits per day.

References:

1. *Nair, Sangeeta. Chandrayaan-2 completes one year on Moon orbit. Jagran Josh. Aug 21, 2020. https://www.jagranjosh.com/current-affairs/chandrayaan-2-completes-one-year-on-moon-orbit-isro-lunar-mission-1598001313-1*

2. *The Indian Express. PTI, Bengaluru. Chandrayaan-2 spacecraft completes over 9,000 orbits around moon, says ISRO. September 6, 2021. https://indianexpress.com/article/cities/bangalore/chandrayaan-2-spacecraft-9000-orbits-moon-isro-7492866/*

ISRO will re-attempt a landing with Chandrayaan-3.

Q244. What did Chandrayaan-2 discover?

A244. Some of the findings of the Chandrayaan 2 are:

1. A key outcome from Chandrayaan-2 has been the exploration of the Permanently Shadowed Regions (PSR's) as well as craters and boulders underneath the Moon dust (regolith), the loose deposit comprising the top surface extending up to 3-4 metres in depth. Chandrayaan-2 pointed to the distribution of most possible locations for the existence of water-ice on the lunar pole. In a major achievement for Chandrayaan-2, one of the eight payloads onboard the orbiter, Dual Frequency Synthetic Aperture Radar (DFSAR), has confirmed the presence of water ice in the permanently shadowed regions on the moon. DFSAR is capable of peering inside the lunar surface up to a few metres' depth.

2. The PSRs in Lunar poles are reported to contain water-ice of various concentrations mixed with the lunar regolith.

3. Chandrayaan 2 is creating a highest resolution map of the moon. Planetary scientists are building the most detailed maps we have ever seen of the moon's surface. It has been able to peer into the permanently shadowed craters at the moon's southern pole and build high-resolution maps of the craters and mountains on the moon's surface.

4. The Terrain Mapping Camera-2 (TMC 2) of Chandrayaan-2, which is conducting imaging of the moon at a global scale, has found interesting geologic signatures of lunar crustal shortening, and identification of volcanic domes

5. The Chandra's Atmospheric Composition Explorer-2 (CHACE-2) onboard Chandrayaan-2 has made the first-of-its-kind discovery on the distribution of Argon-40 to study the lunar exosphere as well as the surface from where this gas is understood to have escaped. Argon-40 originates from the radioactive decay of Potassium-40 present below the lunar surface. Once formed, it diffuses through the inter-granular space in the lunar surface dust and makes its way up to the lunar exosphere through seepage and faults.
6. Chandrayaan-2 has also led to studies around the sun. The spacecraft found abundances of magnesium, aluminium, and silicon in the solar corona and provided new insights about coronal mass heating. Chandrayaan-2 also detected solar proton events due to high intensity solar flares in January 2022.
7. Chandrayaan 2 orbiter's data has continued to generate significant new scientific results, from creating the first ever global-scale sodium maps of our Moon to uniquely characterizing plasma in the lunar exosphere, and even uniquely studying our Sun.

References

1. *Chandrayaan-2 Latest Updates. Department of Space-ISRO. Government of India.*
https://www.isro.gov.in/chandrayaan2-latest-updates

2. *Sibu Tripathi, Sibu. Chandrayaan-2 finds water-ice on the dark side of Moon that has not seen sunlight in two billion years. India Today. September 10, 2021.*
https://www.indiatoday.in/science/story/chandrayaan-2-mission-water-on-moon-dark-side-isro-1850931-2021-09-09

3. *Chandrayaan-2 observes Argon-40 in lunar exosphere. The New Indian Express. March 9, 2022.*
https://www.newindianexpress.com/nation/2022/mar/09/chandrayaan-2-observes-argon-40-in-lunar-exosphere-2427922.html

Q245. Why are we interested in the moon-landings on moon's south pole?

A245. There are several reasons for choosing to land in this location.

1. The South Pole has been identified as a possible future location for a human outpost and the poles are the new frontiers for lunar exploration missions.
2. Recent orbiting missions have shown that the poles may contain water and it is believed that solar wind particles are trapped there. The water and volatile chemicals hold information on the history and chemistry of the inner solar system and could even provide resources for future exploration
3. The South Pole is at the edge of the Aitken basin, the largest impact basin in the solar system. The surface here is made of material from deep within the lunar crust and mantle. Probing these materials could provide insights into planetary formation and reveal its potential for utilization as a resource for future missions.

Q246. What are ISRO's past and future space/lunar exploration missions?

A246. Some of the major completed missions are:

1. **Chandrayaan 1:** Start date October 22, 2008; End date August 28, 2009
2. **Chandrayaan-2:** Start date July 22, 2018; Orbiter still functional
3. **Mangalyaan:** It is India's Mars Orbiter Mission, also known as MOM. It is an interplanetary mission. Start date was November 5, 2013, and still orbiting Mars since September 24, 2014. ISRO has become the fourth space agency to reach Mars, after the Soviet space programme, NASA, and the European Space Agency.
4. **AstroSat (Astronomy Satellite):** Astrosat is India's first dedicated multi-wavelength space telescope. It was launched

on September 28, 2015, into a near-earth, equatorial orbit and is still performing. The five instruments on board cover different wavelengths regions of the electromagnetic spectrum. It is observing various celestial bodies and cosmic sources in multi-wavelengths. With the success of this satellite, ISRO has proposed launching a successor for Astrosat.

5. **RISAT-1A** (EOS-04 or earth Observation Satellite – 04) is a remote sensing satellite with primary application in terrain mapping and analysis of land, ocean, and water surface for soil moisture. It is the sixth in series of RISAT satellites. It is designed to provide high quality images under all weather conditions for applications such as Agriculture, Forestry and Plantations, Soil Moisture and Hydrology and Flood mapping. It was launched on February 14, 2022.

Planned missions

Gaganyaan is an Indian crewed orbital spacecraft intended to be the Indian human spaceflight programme. The objective is to carry a crew of up to three astronauts to low earth orbit (400 km (250 mi)) altitude for up to seven days, perform a set of predefined activities in space, such as rendezvous and docking capability, and return them safely to earth.

As per India Today December 21, 2022, Gaganyaan mission will now launch in the fourth quarter of 2024.

The mission will have the first two test flights Gaganyaan-1 and Gaganyaan-2 prior to the crewed mission Gaganyaan-3 expected to launch no earlier than 2025.

Though the spacecraft is designed to carry 3 people, it is likely that the first flight will carry one person only.

Aditya-L1 is India's space-based astronomy mission to study sun. It will be inserted in a halo orbit around the Lagrange point L1(about 1,500,000 km (930,000 mi) from earth) between earth and sun and will study sun's corona, chromosphere, photosphere, solar winds and flares, solar emissions, and Coronal Mass

Ejections (CMEs). It will also carry out round-the-clock imaging of the sun.

It is planned to be launched in the mid 2023.

Note: The Lagrange points are points near two large orbiting bodies where the gravitational forces and the centrifugal force balance each other. This makes Lagrange points an excellent location for satellites as these can be used by spacecraft to reduce fuel consumption needed to remain in position. There are five Lagrangian points L1 to L5 for the sun–earth system, and in a similar way there are five different Lagrangian points for the earth–moon system.

Chandrayaan-3 was conceived by ISRO after it failed to soft-land the Vikram lander of Chandrayaan-2 on the lunar surface while it still has a fully operational orbiter going around the moon. In a way it will be a mission repeat of Chandrayaan-2 but will only include a lander and rover. It will not have an orbiter. The spacecraft is planned to be launched in the later part of 2023.

As of February 2023, the lander and rocket are undergoing the standard series of space-simulating tests. The lander will attempt a lunar touchdown at the near-polar location, roughly just 100 kilometers east of the Chandrayaan 2 landing site lying in the same larger rocky highland region.

ISRO and NASA successfully tested data flow between India's soon-launching Chandrayaan 3 Moon lander and the US Deep Space Network (DSN). This is part of a February 2022 agreement to have DSN support communications for the mission. ESA's Estrack network will also provide backup communications. ISRO uses its own Indian Deep Space Network as a primary means for communicating with its planetary spacecraft.

NISAR (NASA-ISRO Synthetic Aperture Radar) mission is a joint project between NASA and ISRO to co-develop and launch a dual-frequency synthetic aperture radar on an earth observation satellite. It will spot warning signs of imminent volcanic eruptions, help to monitor groundwater supplies, track the melt rate of ice sheets tied to sea level rise, and observe shifts in the distribution of vegetation around the world. Monitoring these kinds of changes in earth's surface over nearly the entire globe has not been done before with the high resolution in space and time that NISAR will deliver. The expected launch date is January 2024.

Shukrayaan-1 (The Indian Venusian orbiter mission) is ISRO's proposed orbiter to Venus to study the surface and atmosphere of Venus. The three broad research areas of interest include surface/subsurface stratigraphy and re-surfacing processes; study the atmospheric chemistry, dynamics, and compositional variations; and study of solar irradiance and solar wind interaction with Venus's ionosphere. The expected launch date is not before December 2024 and a backup date in 2026 and 2028. As per news dated 17 January 2023 from drishtiias.com, Shukrayaan-1 may be postponed to 2031. It is interesting to note that the optimal launch windows from earth to Venus occur once every 19 months to keep fuel required at liftoff in mind.

As of November 2020, ISRO has shortlisted 20 international proposals that include collaboration with Russia, France, Sweden, and Germany.

Lunar Polar Exploration mission (LUPEX), also known as Chandrayaan-4 is a robotic lunar mission concept by Indian Space Research Organization (ISRO) and Japan Aerospace Exploration Agency (JAXA) that would send a lunar rover and lander to explore the south pole region of the moon. The main objective of the LUPEX Mission is water prospecting and analysis. For this, the mission will demonstrate new surface exploration technologies related to vehicular transport and lunar night survival. In addition, the mission aims for sustainable exploration in the polar region of the moon. JAXA and NASA have discussed the possibility of NASA's participation as well. The proposed launch date is no earlier than 2025.

The ISRO will provide a ~6000-kilogram lander, while JAXA will be responsible for the launch vehicle H3 rocket and a ~350-kilogram rover.

While LUPEX is yet to be formally approved by the Indian government, JAXA and ISRO have finalized the rover instruments, and preliminary design reviews for the mission are underway. The rover will feature not only Indo-Japanese instruments but one contribution each from NASA and ESA as well.

Mangalyaan 2 Mars Orbiter Mission 2 (MOM 2), also called Mangalyaan-2 is India's second interplanetary mission. The mission will consist of only of an orbiter. ISRO plans to launch this mission by 2025. In any case, it will be undertaken only after the launch of Chandrayaan-3.

Space Station: India is planning its first indigenously made Space Station by 2030. According to ISRO's management, the facility would weigh as much as 20 tons, will be fixated at an orbit 400 kilometres above the earth and astronauts could stay for 15-20 days. The space station will be launched 5-7 years after the Gaganyaan mission.

Lunar Colonization: India is planning to build structures on the moon. ISRO, along with academic institutions, is doing experimentation on potential structures. It is at a preliminary stage.

Agreements and Cooperation with Other Countries

Indian government's Department of Space is currently working majorly with nations like Russia, the US, France, Japan, and Israel in the domain of space science and technology. As per Indian Government's 2021-22 report India has signed space cooperation documents with agencies of 60 countries and five multinational bodies. Some of the major areas of cooperation between ISRO and foreign space agencies are:

1. NASA-ISRO Synthetic Aperture Radar (NISAR) mission.
2. Regarding Russia, four Indian astronaut candidates have successfully completed the general astronaut training at Gagarin Cosmonaut Training Centre, Russia.
3. French Space Agency CNES and ISRO are taking their ties beyond Earth observation and working in the areas of Human Spaceflight and Space Situational Awareness (SSA -

the tracking of space debris and orbital objects). In addition to this, there are also deliberations over establishing an ISRO ground station in French Guyana.

4. ISRO and Japanese Space Agency JAXA are working together in the fields of lunar exploration, satellite navigation and Earth observation. The two agencies are working on completing the phase-A study of the Indo-Japan Lunar polar exploration (LUPEX) mission.
5. ISRO and the European Space Agency (ESA) are working on network and operations support for Chandrayaan-3, Aditya L1, etc.
6. As part of India-Australia space cooperation, discussions are progressing towards establishing ISRO's transportable ground station at Australia's Cocos-Keeling Island (CKI) for Gaganyaan mission.
7. ISRO and Israel Space Agency (ISA) have discussed about cooperation in flying ISA's Electric Propulsion System (EPS) in ISRO's small satellite.
8. ISRO and the Italian Space Agency (ASI) have decided to explore cooperation opportunities in earth observation, space science, robotic and human exploration.

Reference

'India has signed documents on space cooperation with 60 countries' - India News News (wionews.com)
Written by Sidharth MP, Updated Apr 28, 2022
https://www.wionews.com/india-news/india-has-signed-documents-on-space-cooperation-with-60-countries-474650

❑

CHAPTER 23

Europe's Lunar Exploration Programme

Q247. Which European agency is involved in the space/lunar exploration activities?

A247. The European Space Agency (ESA), an intergovernmental organization created in 1975, provides Europe's gateway to space. It has 22 member countries: Austria, Belgium, the Czech Republic, Denmark, Estonia, Finland, France, Germany, Greece, Hungary, Ireland, Italy, Luxembourg, the Netherlands, Norway, Poland, Portugal, Romania, Spain, Sweden, Switzerland, and the United Kingdom. In addition, Latvia, Lithuania, and Slovenia are Associate Members. ESA has signed different types of agreements with Canada, Bulgaria, Cyprus and Slovakia, Croatia, and Malta.

ESA's job is to draw up the European space programme and carry it through. Its programmes are designed to find out more about the earth, its immediate space environment, our solar system, and the universe, as well as to develop satellite-based technologies and services, and to promote European industries.

By coordinating the financial and intellectual resources of its members, ESA can undertake programmes and activities far beyond the scope of any single European country. ESA works closely with European industry, national space agencies and the European Union, and cooperates with space agencies worldwide.

Q248. What are the European Space Agency's lunar exploration missions?

A248. The European Space Agency has developed an exploration programme based on four main missions:

1. Luna-Resurs project: It is a partnership project with the Russian agency Roscosmos. The project will carry European technology to land precisely and safely on the moon and to extract and analyse samples of the lunar terrain.

   Note: UKRAIN WAR: In March/April 2022, The European Space Agency (ESA) said that the ESA will "discontinue cooperative activities" with the Russian space agency Roscosmos citing Russian aggression against Ukraine and sanctions placed on Russia in its reasoning.

2. Service module to Orion spacecraft: ESA will provide the service module to NASA's Orion spacecraft (a class of partially reusable crewed spacecraft to be used in NASA's Artemis programme). ESA's service module will provide propulsion, life support, power, air, and water as well as control the temperature in the crew module.

3. In-Situ Resource Utilization (ISRU): One of the biggest challenges, when on another celestial body, involves the extraction and processing of local resources into useful products and services, known as in-situ resource utilization. Missions to explore the lunar natural resource potential are foreseen by ESA by 2025. The goal is to produce drinkable water or breathable oxygen on the moon.

4. The Heracles mission: ESA is working with the Canadian and Japanese space agencies to prepare the Heracles robotic

mission to the moon. It could take off in 2028 to allow ESA to gain knowledge on human-robotic interaction while landing a spacecraft on the moon, collecting samples with a rover operated from the lunar Gateway and sending samples back to earth.

Note: The "Lunar Gateway" will be a multi-purpose outpost orbiting the moon. It will provide essential support for long-term human return to the moon and will serve as a staging point for deep space exploration. NASA is working with commercial and international partners to establish the Gateway. It will be a part of NASA's Artemis project.

5. The ESA (EU) is also working on a new system called "LunaNet" to handle lunar communication and navigation services which might result in the Moon getting its time zone. When humans return to the Moon, they will have a challenge in figuring out what time it is. Since astronauts on the Moon aren't on Earth, they won't be a part of a specific time zone. This sounds like a trivial problem, but when you think of all the computers measuring and communicating time with each other, getting the time wrong could be a risk to astronaut safety.

Q249. What was the European Space Agency's first lunar probe?

A249. SMART-1 (Small Missions for Advanced Research and Technology-1) was the first lunar probe of the European Space Agency, designed and developed by the Swedish Space Corporation. It was a small (about one metre across) lightweight probe launched on September 27, 2003. Its primary mission goal was to demonstrate Solar Electric Primary Propulsion by orbit raising from earth's geostationary transfer orbit (GTO) to the moon.

It travelled to the moon using solar-electric propulsion and carrying a battery of miniaturized instruments. SMART-1 did the first comprehensive inventory of key chemical elements in the lunar surface. It scanned the moon for signs of water in polar craters. It also investigated the theory of the moon's formation by a violent collision of Theia with earth. When its mission ended, it crashed into the moon on September 3, 2006. Telescopes on earth observed the impact to study the lunar surface.

Notes:

1. SMART is a series of small, relatively low-cost ESA missions to test new technologies that could be used on bigger projects.
2. Solar electric propulsion combines solar panels on spacecraft and electric thrusters to propel a spacecraft through outer space. This technology has been exploited in a variety of spacecraft by ESA, Japan's JAXA, India's ISRO, and NASA.
3. Geostationary Transfer Orbit (GTO): GTO is a type of geocentric orbit. Satellites that are destined for geosynchronous orbit (GSO) or geostationary orbit, also referred to as a geosynchronous equatorial orbit (GEO) are (almost) always put into a GTO as an intermediate step for reaching their final orbit. A GTO is highly elliptic.

Objects in GSO have an orbital speed that matches the earth's rotation, yielding a consistent position over a single longitude. GEO is a kind of GSO. It matches the planet's rotation, but GEO objects only orbit earth's equator, and from the ground perspective, they appear in a fixed position in the sky.

Q250. Why does the European Space Agency's SMART-1 represent a breakthrough in planetary exploration?

A250. SMART-1 was the first European spacecraft to use an ion engine as a main propulsion system (Earlier NASA used it in Deep Space 1 probe launched in October 1998). It also introduced an additional challenge. To go to the moon, the spacecraft combined for the first time ever solar electric propulsion (EP) with maneuvers using the gravity of celestial bodies. Engineers considered the gravity of earth and the moon at each stage of the journey and adjusted accordingly the thrust provided by the ion engine. Flight experience using EP was still relatively new, and SMART-1 was the first mission to use EP to achieve large-scale orbit raising from around the earth. On November 11, 2004, SMART-1 passed into the primary influence of the moon and made history with several notable firsts including:

(1) Being the first Electric Propulsion mission to escape earth orbit

(2) The first to use Electric Propulsion to enter into orbit around another celestial body, and

(3) Europe's first Lunar mission.

> Note: An ion engine is a form of electric propulsion used for spacecraft propulsion. It creates thrust by accelerating ions using electricity.

ESA's SMART-1- testing solar electric propulsion and studying the moon
Credits: Wikimedia Commons, ESA
https://commons.wikimedia.org/wiki/File:ESA%27s_SMART-1-_testing_solar_electric_propulsion_and_studying_the_moon_ESA194476.jpg
Author: ESA. Date April 23, 2002
Source: http://www.esa.int/spaceinimages/Images/2002/04/ESA_s_SMART-1_testing_solar_electric_propulsion_and_studying_the_moon
File is free to share – to copy, distribute and transmit the work

Q251. What is the European Space Agency's SpaceBok robot?

A251. The European Space Agency is currently testing a prototype quadrupedal robot called SpaceBok, a hopping robot built to leap across the surface of low-gravity celestial bodies. SpaceBok is designed by a team from ETH Zurich and ZHAW Zurich for future missions to the moon and Mars. With a hip height of 500 mm (19.7 in) and a mass of 22 kg (48.5 pounds), its dimensions are comparable to a medium-sized dog. In contrast to other legged robots, SpaceBok is primarily built for hopping. This would allow for a fast and efficient way of moving forward by overcoming obstacles on low-gravity celestial bodies. Based on test experiments and simulation results, conducted by the team, the robot can move with up to 1.1 m/s (3.6 ft/sec) for low gravity on the moon. The robot achieves repetitive, vertical jumps of more than 0.9 m (3 ft) and powerful single leaps of up to 1.3m (4.3 ft). This would allow it to travel in the difficult terrain and could be used to explore areas of interest, such as craters, which most other rovers are unable to reach.

SpaceBok robot
Credits: Robotic Systems Lab/ETH Zurich
Released date November 28, 2018
https://rsl.ethz.ch/robots-media/spacebok.html
Reproduced with the permission

References

1. *Arm, Philip. Zenkl, Radek. et al. SpaceBok: A Dynamic Legged Robot for Space Exploration. Published in 2019 International Conference on Robotics and Automation (ICRA) Date of conference May 20-24, 2019.* https://www.research-collection.ethz.ch/handle/20.500.11850/384203
2. *Kolvenbach, Hendrik. Hampp, Elias et al. Towards Jumping Locomotion for Quadruped Robots on the Moon. Published in 2019 IEEE/RSJ International Conference on Intelligent Robots and Systems (IROS). Date of Conference: November 3-8, 2019.* https://www.research-collection.ethz.ch/handle/20.500.11850/438204

❑

CHAPTER 24

Japan's Lunar Exploration Programme

Q252. What was Japan's Hiten programme?

A252. Hiten, originally known as Muses-A (Mu Space Engineering Satellite-A), was a technology demonstration satellite. It was the first Japanese lunar probe sent to the moon. It was also the first lunar probe launched by a country other than the Soviet Union or the United States. The probe was launched on January 24, 1990. On the first lunar approach, Hiten released a small orbiter Hagoromo into lunar orbit. It successfully demonstrated the aerobraking technique on March 19, 1991. This was the first aerobraking manoeuvre by a deep space probe. The primary mission of the probe was concluded after the second aerobraking manoeuvre on March 30, 1991. These technological data are used in spacecraft development.

> **Note:** Aerobraking is a spaceflight technique wherein an orbiting spacecraft brushes against the top of a planetary atmosphere. The friction of the atmosphere against the surface of the spacecraft slows down and lowers the craft's orbital altitude.

Q253. What is Japan's Kaguya lunar exploration mission?

A253. Kaguya was Japan's second unmanned mission to the moon. It was launched by the Japan Aerospace Exploration Agency (JAXA) in September 2007. Its prelaunch name was SELENE (Selenological and Engineering Explorer). Kaguya comes from the name of a lunar princess in the ancient Japanese folktale. It was the largest lunar mission since the Apollo programme.

The main scientific objectives of the mission were to study the origin of the moon, obtain information about the lunar surface environment, measurement of the moon's gravity field, and to develop the technology for future lunar exploration.

In addition to the main orbiter, Kaguya, the mission included two small sub-satellites named Okina and Ouna which were used to ensure continuous communication between earth and Selene and to help map gravity variations in the moon as mass variations accelerated and decelerated the craft. On October 3, 2007, Kaguya entered into an initial elliptical polar orbit around the moon, the first time that a Japanese spacecraft had done so. Okina and then Ouna were released into elliptical polar orbits on October 9 and 12, 2007. The main spacecraft then entered a circular polar orbit by October 19. Okina crashed into the moon on Feb. 12, 2009. Selene crashed into the moon on June 10, 2009.

Major achievements include:

1. By April 9, 2008, Kaguya had collected enough data to construct the topography of the entire lunar surface, with data points 10 orders larger than the previous model of the lunar surface.
2. Detecting gravity anomalies on both the near and the far side of the moon and generating detailed gravity map of the far side of the moon. The moon has a much "lumpier" gravity field than earth because its mass is not as smoothly distributed inside
3. Making the first optical observation of the permanently shadowed interior of the crater Shackleton at the lunar south pole, and

Three-dimensional model of Japanese space probe Kaguya (known also as Selenological and Engineering Explorer and SELENE), launched on September 14, 2007 and developed for the exploration of the moon.
Credits: Wikimedia Commons
https://commons.wikimedia.org/wiki/File:Selene.gif
Author: NASA
Source: NASA Space Science Data Coordinated Archive (NSSDCA)
Figure in Public domain

4. Finding evidence of the earth's oxygen being transported to the moon via magnetospheric ions.

Q254. What are Japan's lunar exploration projects?

A254. Japan has an ambitious space/moon exploration programme, which is briefly outlined below.

1. Smart Lander for Investigating Moon (SLIM)

SLIM is a lunar lander being developed by the Japan Aerospace Exploration Agency (JAXA). It will be Japan's first lunar surface mission. It aims to demonstrate "landing where it is desire to land", pin-point landing technique and obstacle detection technique.

The following are the SLIM mission objectives:

1. Demonstration of the accurate lunar landing techniques embodied in a small explorer
2. Acceleration of the study of the moon and other planets using the lighter exploration system

The landed weight will be about 200 kg (440 pounds) and the landing objective is to be within 100 metres of the target point.

It is now proposed to be launched in August 2023 timeframe. After landing, the SLIM project is planning to observe the lunar surface around landing site using the Multi-Band Camera (MBC). By succeeding in this extremely precise landing, it will lead to enhancing the quality of space exploration.

2. Japan's Transformable Robot Ball

Japan Aerospace Exploration Agency (JAXA) has developed an unusual robot to explore the Moon. Its size is like a baseball with 3 inches (80mm) diameter will weighs about 250 grams (0.5 pounds). Japan's ispace’s Hakuto-R Mission 1 lander carried it to the lunar orbit, but the lander failed to soft land on the Moon’s surface on April 25, 2023. While on the moon, the robot was supposed to open and roll around on its two wheels on the lunar surface, take pictures of the moon's surface and its dust (regolith) and then send them back to mission control for analysis.

> **Note:** Japan’s ispace will not only be transporting the transforming robot ball of JAXA but it has also been selected to transport rovers from Other Countries like Canada and UAE to the moon.

3. OMOTENASHI

The Japanese OMOTENASHI (Outstanding MOon Exploration TEchnologies demonstrated by NAno Semi-Hard Impactor), the world's smallest moon lander, aimed to demonstrate a low-cost technology semi-soft landing to land and explore the lunar surface with a 12 kg (26.5 pounds) spacecraft. It was a 6U CubeSat set with a mission period of 4 to 5 days.

OMOTENASHI was launched with the Artemis 1 mission via a NASA SLS rocket on 16 November 2022. After deployment from the Artemis I, JAXA reported unstable communications with the spacecraft. On 21 November 2022, JAXA reported that further attempts to communicate with the lander had ended.

Omotenashi is a Japanese word for "welcome" or "Hospitality".

Notes:

1. A CubeSat (U-class spacecraft) is a type of miniaturized satellite (known as nanosatellites) for space research that is made up of multiple cubic modules of 10 cm × 10 cm × 10 cm size. CubeSats have a mass of no more than 1.33 kilograms (2.9 pounds) per unit. CubeSats are put into orbit by deployers on the International Space Station or launched as secondary payloads on a launch vehicle.

2. NASA's SLS (Space Launch System) rocket is a super-heavy lift launch vehicle that provides the foundation for human exploration beyond earth's orbit. With its unprecedented power and capabilities, SLS is the only rocket that can send Orion (a crewed spacecraft), astronauts, and cargo to the moon on a single mission.

4. EQUULEUS

Japan's EQUULEUS (EQUilibriUm Lunar-earth point 6U Spacecraft) flew to the earth-moon Lagrangian Point L2 (where the earth's and moon's gravitational pulls roughly balance) to study earth's plasmasphere. It will also measure meteor impact flashes and the dust environment around the moon, providing additional important information for safe human exploration. EQUULEUS is a 6U CubeSat with dimensions 10 x 20 x 30 cm and a mass of approximately 12 kg (26.5 pounds).

It Was launched by NASA's Artemis I on November 16, 2022. As a result of the orbit maneuver control and orbit corrections before and after the lunar fly-by, the lunar fly-by was completed as planned on November 22 and EQUULEUS has been confirmed to be in the planned orbit. This is the world's first successful orbit control beyond low-Earth orbit using a water propellant propulsion system. From now, EQUULEUS will move from the initial operation phase onto the normal operation phase and reach the Lagrange point in approximately 1.5 years.

5. Japan aims to put a person on the moon by late 2020s

Japan is aiming to put a Japanese person on the moon by the latter half of the 2020s. The mission is a part of a US-led initiative, the Artemis programme, that aims to return astronauts to the moon. The plan also spells out Japan's aspirations to launch a probe to explore Mars in 2024, as well as to find ways to generate solar electricity in space.

It Was launched by NASA's Artemis I on November 16, 2022. As a result of the orbit maneuver control and orbit corrections before and after the lunar fly-by, the lunar fly-by was completed as planned on November 22 and EQUULEUS has been confirmed to be in the planned orbit. This is the world's first successful orbit control beyond low-Earth orbit using a water propellant propulsion system. From now, EQUULEUS will move from the initial operation phase onto the normal operation phase and reach the Lagrange point in approximately 1.5 years.

6. JAXA and Toyota's Lunar Cruiser

JAXA and Toyota have agreed to build a six-wheeled, self-driving fuel cell electric vehicle (named Lunar Cruiser) to explore the surface of the moon. Japan is hoping to launch the vehicle to the moon as soon as 2029. The rover, roughly 2 microbus size, will be able to carry two passengers (or four in an emergency) for more than 10,000 km (6,200 miles) across the lunar surface. Power will come from an array of replaceable fuel cells which will be charged over time using foldable solar panels.

7. Japan's ispace Hakuto-R Mission 1 lunar lander

Japanese ispace, inc., ("ispace"), ispace, inc., ("ispace") a private global lunar exploration company, has built the HAKUTO-R Mission 1 lunar lander (also known as M1 lander). The M1 lander was expected to deploy a two-wheeled, baseball-sized rover from Japan's JAXA space agency and a four-wheeled Rashid rover made by the United Arab Emirates. It was launched on December 11, 2022, on a SpaceX Falcon 9 rocket from Cape Canaveral, Florida. It was on track for a lunar landing attempt in late April 2023 within the Atlas crater. On January 20, 2023, the spacecraft reached its farthest point from Earth some 8.5 million miles away (1.375 million km). According to a statement dated April 15, 2023, from ispace, the lander was in orbit around the Moon targeting a landing. It tried to land on April 25, 2023, but failed to soft land.

As a follow up, Ispace are developing Mission 2 and Mission 3 concurrently.

The name HAKUTO refers to the white rabbit that lives on the Moon as per Japanese folklore.

❑

CHAPTER 25

Lunar Exploration Programme of Other Countries

(Australia, Canada, Emirates, Israel, South Korea, Turkey, and UK's Spacebit)

Q255. What is the Australia's lunar mission?

A255. The Australian Space Agency has announced to send an Australian made rover to the moon by 2026 under a deal with NASA. The rover will collect lunar soil containing oxygen, which could eventually be used to support human life in space. The rover, weighing less than 20 kilograms, will go to the moon onboard a commercial lunar lander funded by NASA. The rover will explore the moon's south pole and collect lunar soil to transfer it to the lander where a NASA-operated system will extract oxygen from it. This mission will thus demonstrate a key step towards establishing a sustainable human presence on the moon.

A separate mission, jointly planned by private organizations/ university of Australia, Japan, and Canada, may see Australian technology hunting water on the moon as soon as mid-2024. According to this plan, the ten-kilogram rover, measuring

60×60×50 cm (2×2×1.64 ft), will be launched on board Japan's company ispace. The rover, also built by ispace, will have an integrated robotic arm created by Canada and Australia. Using cameras and sensors, the arm will collect high-resolution data and will also collect information on the physical and chemical composition of lunar dust, soil, and rocks, specifically with a goal of finding water.

Q256. What is Canada's role in lunar exploration activities?

A256. The Canadian Space Agency (CSA) is the national space agency of Canada, established in 1990. The agency is responsible to the minister of innovation, science, and economic development.

Canada has officially announced its participation in the exciting next chapter of moon exploration, which will include:

1. A smart robotic system known as Canadarm3 for the "Lunar Gateway", a small space station that will orbit the moon. Canadarm3 will be comprised of a large and a small robotic arm, specialized tools to perform maintenance and science tasks, and robotic interfaces to enable access to all Gateway elements. It will be highly autonomous and will use its advanced AI-enabled sensors and systems to make decisions when the astronauts are not present.

In return for contributing Canadarm3, Canada receives a range of opportunities for lunar science, technology demonstration and commercial activities, as well as two astronaut flights to the moon. A CSA astronaut will be part of NASA's Artemis II, the first crewed mission to the moon since 1972.

2. A programme for innovative Canadian science and technology, including a Canadian rover designed for the moon. The CSA launched a new initiative known as the Lunar Exploration Accelerator Programme (LEAP). This programme prepares Canada's space sector for lunar exploration by offering a wide range of opportunities for Canadian science and technology activities in lunar orbit or on the moon's surface.

One of the programme elements within LEAP is the development, launch, and operation of a Canadian rover, set to perform technology demonstration. It will include one or more Canadian science instruments, on the surface of the moon. The Canadian rover will explore a polar region of the moon within the next five years. The mission will aim to gather imagery, measurements, and data on the surface of the moon, as well as to have the rover survive an entire night on the moon. Lunar nights, which last about 14 earth days, are extremely cold and dark, posing a significant technological challenge.

Q257. What is the Emirates' lunar mission?

A257. The Emirates Lunar Mission is the first mission to the moon from the United Arab Emirates (UAE). The mission by Mohammed Bin Rashid Space Centre (MBRSC) has sent a lunar rover named Rashid to the moon aboard Japan's ispace's Hakuto-R Mission 1 lander. Rashid had six instruments and weighed about 10 kg. It is one of the smallest rovers to land on the moon. It was launched on 11 December 2022 and tried to soft land on the Moon on April 25, 2023 but failed to do so.

The rover was planned to study:

1. the lunar surface, mobility on the moon's surface and how different surfaces interact with lunar particles.
2. explaining why moon dust is so sticky.
3. deciding the materials used in spacesuits and the landing mechanisms used to land humans on the moon.

UAE's space agency have revealed that they plan to send another rover to the moon before 2025, followed by several orbiters. They will also look to partner with other agencies—presumably NASA—to send Emirati astronauts to the moon.

Q258. What are Israel's lunar exploration activities?

A258. The Israel Space Agency (ISA) coordinates all Israeli space research programmes. It was founded in 1983 and is a part of Israel's Ministry of Science and Technology. ISA is one of only seven countries that both build their own satellites and launch their own launchers.

ISA's Moon Exploration Activities

(1) ISA has been collaborating with NASA about future lunar research programmes. It has established the "Israel Network for Lunar Science and Exploration (INLSE)" programme as a part of an international effort to study the moon and the solar system. The primary goal is to train personnel in lunar science and exploration and to build the required infrastructure to collaborate with other nations in these fields. INLSE, through its technical and engineering expertise, focuses on laser communication, robotic, remote sensing, and other technologies for lunar missions as areas for possible collaboration both within Israel and through the Solar System Exploration Research Virtual Institute (SSERVI).

(2) ISA has sponsored the first Israeli effort to land a spacecraft on the moon. An Israeli organization, SpaceIL, established in 2011, competed in the Google Lunar X Prize (GLXP) contest to land a spacecraft on the moon. SpaceIL is a non-governmental organization made up of multidisciplinary team of Israeli scientists and astronomers. The team developed a robotic spacecraft Beresheet (Beresheet means "In the Beginning" in Hebrew) weighing around 500 kg. It was designed to be launched and then land on the moon bearing the Israeli flag. It was Israel's first lunar mission and the first attempt by a private company to land on the moon. The mission achieved lunar orbit but was lost during an April 2019 landing attempt. Israel has announced that a new mission, named Beresheet 2 would attempt a second time to land on the moon. The planned date is 2025.

(3) Israel and the UAE cooperation: At World Space Week in Dubai, in October 2021, Israel and the UAE ministers of

science and technology announced plans for cooperation on the lunar exploration mission.

Note: Google Lunar X Prize (GLXP) was a competition sponsored by Google that challenged private companies to land a spacecraft on the moon. Though no company was able to meet the competition deadline, prompting Google to end it with no winner in March 2018, the Israeli team pressed on.

Q259. What is South Korea's lunar exploration Programme?

A259. South Korea's space agency, called Korea Aerospace Research Institute (KARI), has outlined a Korean Lunar Exploration Programme (KLEP) meant to develop and enhance South Korea's technological capabilities, as well as map natural resources from orbit. The programme is divided in two phases:

Phase 1: Launching and operating the Korea Pathfinder Lunar Orbiter (KPLO) named Danuri.

Phase 2: Launching a lunar orbiter, a lander, and a 20 kg rover together on a KSLV-2 South Korean launch vehicle before 2030.

Phase 1: KPLO

KPLO is the first lunar probe by South Korea to orbit the moon with following specific technological objectives:

- Development of critical technologies for lunar exploration
- Produce a topographic map to select future lunar landing sites
- To survey lunar resources such as water ice, uranium, helium-3, silicon, and aluminum
- Development and validation of new space technologies

The spacecraft has a cubic shape with a total mass of 550 kg carrying four South Korean and one US built instrument.

KPLO was launched on 4 August 2022 on a SpaceX Falcon 9 rocket to orbit the Moon for 1 year. The spacecraft reached lunar orbit in December 2022 and began its scientific mission in February 2023. It will analyze and map the lunar surface, measure gamma rays and magnetic strength, and test an experimental "space internet" technology to send images of the Moon home to Earth. It has sent Home Pictures of the Earth and Moon.

In March 2023, KPLO's highest resolution imager photographed several pre-identified places. South Korea will use images to help identify good landing sites for its fully indigenous Moon lander (targeting launch in 2032). KPLO's wide-angle polarimetric camera has captured polarization data of various areas. This capability will be helpful to understand the origin(s) Moon- swirls. The orbiter's gamma-ray spectrometer and magnetometer have begun taking measurements as well.

KARI plans to release a bunch of datasets from all these four instruments by the end of April 2023.

Phase 2

South Korea's President has vowed to launch the nation's first robotic lunar lander on a domestically developed rocket by 2032.

Q260. What is Turkey's lunar exploration programme?

A260. The Turkish Space Agency (TUA), established in 2018, is the government agency for national aerospace research in Turkey. Country's 10-year master plan includes missions to the moon, sending Turkish astronauts into space, and developing internationally viable satellite systems.

The first stage of the mission would be through international cooperation, while the second stage would utilize Turkish rockets. As per the timeline, TUA is planning to send a robotic spacecraft to the moon towards the end of 2023 (when the country marks the centennial of the founding of the Turkish republic) with international cooperation. They are in talks with Russia and SpaceX for a possible launch. The second phase of the program is planned to be carried out in 2028. Turkey also plans to soft land a rover to the moon by 2030 using a domestically built rocket engine.

Turkey's Space Launch System project will develop country's satellite launch capability. The aim of the project is to support the sustainability of the national satellite programmes and to reach space independently.

The last objective is planned for the second half of 2023 by sending a Turkish citizen to the International Space Station (ISS) within the scope of the Turkish Space Traveler and Science Mission project.

Turkey planned to establish "a first contact with the moon" in 2023.

Q261. What is Spacebit's (UK) walking robot Asagumo?

A261. British robotics company, Spacebit, has created the world's first spider like walking rover named Asagumo. Instead of wheels, Spacebit chose legs. This will allow Asagumo to walk across bumpy terrain and explore lunar lava tubes. Asagumo will be able to explore areas of the moon never seen before.

This first-of-a-kind walking lunar rover stands at only 10 cm (3.9 inches) and weighs 1.3 kilograms (2.9 pounds). It is the world's smallest robotic four-legged lunar rover, designed to be solar powered. It has instruments capable of measuring radiation, taking 3D scanner, filming HD video, and a range of other measurements.

Asagumo will travel up to the moon in the ULA Vulcan launch vehicle being developed by the United Launch Alliance (ULA) since 2014 with an initial flight expected in 2023 and will be delivered to the moon's surface by Astrobotic's Peregrine lunar lander spacecraft.

Asagumo will walk for at least 10 metres on the lunar surface, sending data back to earth.

Planned missions

The first mission will be a tech demo of Asagumo's capabilities on the moon.

In the second mission, estimated for 2023, Spacebit will send a wheeled rover, acting as a mothership which carries 4 Asagumo's inside of it. It will drop these rovers onto the surface to explore more of the surface and transmit data back to earth.

The third mission in 2026 will begin full lunar cave exploration. The rovers will create detailed 3D scans of the cave insides, exploring the tube networks across the moon's surface.

In the next decade, there is hope that Spacebit's technology could contribute to unprecedented exploration on Mars.

> Note: Lunar lava tubes are lava tubes on the moon formed during the eruption of basaltic lava flows. When the surface of a lava flow cools, it hardens, and the lava can channel beneath the surface in a tube-shaped passage. Once the flow of lava diminishes, the tube may drain, forming a hollow void.

Skynews-spacebit-rover-moon
Credits: Spacebit, UK
Wikimedia Commons
https://commons.wikimedia.org/wiki/File:Skynews-spacebit-rover-moon.png
Author: Serhii Harbaruk
Source: Own work
Date January 31, 2020
Free to share, to copy, distribute and transmit the work

❑

PART IV: OTHER TOPICS

"It is a beautiful and delightful sight to behold the body of the moon"

- Galileo Galilei

CHAPTER 26

Other Topics

1. Legal Issues

Q262. What is the Space Law?

A262. It is evident that with the growing space activities, space will become ever more like an extension of earth - an arena for firms and private individuals, not just governments. Therefore, a system of space laws has been created to govern the heavens—both in peacetime and, should it come to that, in war.

Space law is the body of law governing space-related activities, encompassing both international and domestic agreements, rules, and principles. It addresses a variety of matters, such as the preservation of the space and earth environment, liability for damages caused by space objects, the settlement of disputes, the rescue of astronauts, the sharing of information about potential dangers in outer space, the use of space-related technologies, and international cooperation. It is associated with the rules, principles and standards of international law appearing in the five international treaties and five sets of principles governing outer space which have been developed under the auspices of the United Nations.

Note: The Committee on the Peaceful Uses of Outer Space (COPUOS) was set up by the United Nations General Assembly in 1959 to govern the exploration and use of space for the benefit of all humanity: for peace, security, and development.

Q263. What are the five international treaties governing outer space?

A263. The United Nations' Committee on the Peaceful Uses of Outer Space (COPUOS) is the forum for the development of international space law. The Committee has concluded five international treaties and five sets of principles on space-related activities. Each of the treaties stresses the notion that outer space, the activities carried out in outer space and whatever benefits might be accrued from outer space should be devoted to enhancing the well-being of all countries and humankind, with an emphasis on promoting international cooperation.

The treaties commonly referred to as the "five United Nations treaties on outer space" are:

(a). The "Outer Space Treaty"- entered into force on October 10, 1967

Treaty on Principles Governing the Activities of States in the Exploration and Use of Outer Space, including the moon and Other Celestial Bodies

The Outer Space Treaty is the most foundational legal instrument of space law and its broader principles of promoting the civil and peaceful use of space continue to guide multilateral activities in space, such as the International Space Station and the Artemis Programme.

As of February 2021, 111 countries are parties to the treaty. The Treaty spurred to prohibit the use of outer space for military purposes. Key provisions of the Outer Space Treaty include:

- Space is common ground, and everyone is allowed to explore it peacefully. That means no military bases, no nuclear weapons in space. And it's not just military bases — no country can claim any land and any celestial object in space.

- Anything that launches into space must be registered.
- If there's a spaceship crash, the state is held responsible for the damage. There is a separate liability treaty.
- No one is supposed to contaminate space.

(b). The "Rescue Agreement"- entered into force on December 3, 1968

Agreement on the Rescue of Astronauts, the Return of Astronauts and the Return of Objects Launched into Outer Space

States shall take all possible steps to rescue and assist astronauts in distress and promptly return them to the launching state

(c). The "Liability Convention"- entered into force on September 1, 1972

Convention on International Liability for Damage Caused by Space Objects

(d). The "Registration Convention"- entered into force on September 15, 1976

Convention on Registration of Objects Launched into Outer Space

(e). The "Moon Agreement"- (also known as the moon Treaty) entered into force on July 11, 1984.

Agreement Governing the Activities of States on the Moon and Other Celestial Bodies (including the orbits around such bodies)

The Agreement reaffirms and elaborates on many of the provisions of the Outer Space Treaty as applied to the moon and other celestial bodies, providing that those bodies should be used exclusively for peaceful purposes, that their environments should not be disrupted, that the United Nations should be informed of the location and purpose of any station established on those bodies. In addition, the Agreement provides that the moon and its natural resources are the common heritage of mankind and that an international regime should be established to govern the exploitation of such resources when such exploitation is about to become feasible.

As of January 2019, 18 states are parties to the treaty. As of September 2021, it has not been ratified by any state that engages in self-launched human spaceflight (for example, the United States, Russia (former Soviet Union), People's Republic of China) and thus it has little to no relevancy in international law.

Q264. What are the five sets of principles governing outer space?

A264. The five declarations and legal principles are:

(a). The "Declaration of Legal Principles"- December 13, 1963

Governing the Activities of States in the Exploration and Uses of Outer Space.

(b). The "Broadcasting Principles"- December 10, 1982

The Principles Governing the Use by States of Artificial earth Satellites for International Direct Television Broadcasting

(c). The "Remote Sensing Principles"- December 3, 1986

The Principles Relating to Remote Sensing of the earth from Outer Space

(d). The "Nuclear Power Sources Principles"- December 14, 1992

The Principles Relevant to the Use of Nuclear Power Sources in Outer Space

(e). The "Benefits Declaration"- December 13, 1996

The Declaration on International Cooperation in the Exploration and Use of Outer Space for the Benefit and in the Interest of All States, Taking into Particular Account the Needs of Developing Countries

NOTE: It is no secret that the commercial space sector hopes to create lunar tourism and lunar mining, extracting, and selling some of the moon's vast resources on the open market. But there is a snag! According to an international team of scientists led by the Harvard and Smithsonian Centre for Astrophysics, there may not be enough resources on the moon to go around. Without some clear international policies and agreements in place to determine who can claim what and where, the moon could quickly become overcrowded, overburdened, and stripped of its resources.

References

1. *United Nation's Compilation Booklet "International Space Law: United Nations Instruments" Year 2017 (ST/SPACE/61/Rev.2).*
2. *Space Law Treaties and Principles. United Nations – Office for Outer Space Affairs.* https://www.unoosa.org/oosa/en/ourwork/spacelaw/treaties.html

Q265. Can you buy a part of the moon?

A265. The international treaties say that no nation owns the moon, the moon is there for the good of all humans, and so it is impossible for any state to own any lunar land. However, there are certain loopholes in the legal code. Some believe that it is possible to purchase lunar land. Lunar Embassy (https://lunarembassy.com), The Lunar Registry (https://lunarregistry.com), Lunar Land (https://lunarland.com), and Rajat Rajan Saxena of India are some of the individuals/organizations in the business of selling extraterrestrial real estate, including the moon. Many individuals, including celebrities have reportedly bought lunar plots hoping to snatch their share of land before private companies or nations do.

To some people it appears that there is nothing that expressly forbids individuals and organizations from owning land on the moon. However, with no way to claim that land, anyone attempting to sell land to prospective buyers does not make a legal sense. Any documentation that claims you own land on the moon is unenforceable, and no nation that has signed either the Outer Space Treaty or the moon Treaty will recognize it.

With multiple space agencies looking to create "international villages" and companies hoping to create a tourist industry, you could expect some serious legal battles down the road! To this end, a group of legal scholars and space experts recently came together to form the Space Court Foundation (SCF), a non-profit educational organization created to foster a conversation about these and other related space issues. The SCF hopes to play a pivotal role in the evolving domain of space law, which is going to become very important in the coming years!

References

1. *Extraterrestrial real estate*
From Wikipedia, the free encyclopedia
https://en.wikipedia.org/wiki/Extraterrestrial_real_estate

2. *Are People Buying Real Estate on the Moon: Here's What You Need To Know. Lakeside Little Company Blog. August 28, 2020.*
https://lakesidetitle.com/2821-2/

3. *Viegas, Vanessa. Indians are buying plots on the moon, and the reasons are strangely familiar. Hindustan Times News. Updated on Jun 18, 2021.*
https://www.hindustantimes.com/lifestyle/art-culture/indians-are-buying-plots-on-the-moon-and-the-reasons-are-strangely-familiar-101624034427905.html

4. *Williams, Matt. The Space Court Foundation is Now in Session! Universe Today. September 18, 2020.*
https://www.universetoday.com/147832/the-space-court-foundation-is-now-in-session/

2. What if the moon was not there?

Q266. What would have happened if we did not have the moon?

A266. This is a hypothetical question, but an interesting one.

The moon not only lights up the night sky but is also good for a whole lot more. It has played a pivotal role in the development of life on the earth. There will be shorter days, very different seasons, a more extreme climate and much more if there is no moon.

Let us understand one by one!

(a). Tidal Changes

One of the most obvious fallouts without the moon would be the changes in earth's tides. Tides on earth without the moon would be due to the sun's gravitational force only and about a third the size they are now. This would have some very large impacts on life on the earth, such as:

A Day on the earth Would Be Much Shorter; Only About 6-to-8 Hours

The collision of Theia and earth, which formed the moon set earth spinning at a faster rate, (which means shorter days). Geologists have calculated that at the time of its formation, the earth was rotating four times faster than it is today. That means an earth's day was only about 6 hours long and a year would consist of over 1000 days. But due to tidal friction force between the moon and the earth, its rotation has slowed down. The movement of the bulge of tidal water across the oceans and its attraction to the moon acts as a brake on the earth's rotation. Slowly but surely, the length of a day is increasing. The day is currently getting longer by about 0.002 seconds per century. It doesn't sound like much, but over billions of years it adds up. Together with the sun, the moon has caused the day to lengthen from 6 hours to 24. If the moon was not there, this slowing down rate of earth's rotation would have been much less.

Also, at a high spinning rate the earth would have a larger equatorial bulge and much more flattened pole.

The Winds Would be Much Stronger

The faster a planet rotates, the faster its winds blow. Jupiter rotates every 10 hours, and its winds speeds are typically between160 to 320 km/hr (100 to 200 miles per hour). In the moon's absence earth will spin faster, and that would result in stronger winds, and higher and more violent ocean waves. Beach erosion will be more severe, and hurricanes would be much stronger and more devastating. Wind and rain would wear down mountains much faster; the highest mountain peaks would be lower than those on earth today.

Evolution of Life Would Have Been Different

To understand how life evolved, let us go back in time. In the early days, the moon was much closer to the earth and the tides on the young earth were humongous. These tides plunged miles inland and withdrew every three hours (remember, the day was only six hours long). Every time the tides retreated, it dragged back a lot of rocky material from the land into the sea or into small tide pools. Continually churned up in the water, these chemicals formed the complex organic broth in which the first life probably took its shape with sunrays acting as a catalyst.

Take the moon away from the earth and the severity of tides get reduced by a third. Perhaps, that would also slow down the process of forming the chemical building blocks of life. Additionally, stronger winds and violent incessant sea waves would act as a spoiler in the entire process. If life indeed springs on earth, it will probably have a different biology. Plants and animals would remain in the seas for a longer time. High winds would bend and uproot taller plants. Plants would grow close to the ground, develop wide, heavy stems, and anchor themselves with deep roots. Animals would develop wide bodies with short legs. Most would grow thick skins or shells. Wind disperses sound, making it harder to hear. Animals would develop a very keen sense of hearing. Intelligent life might not exist yet. And when they arrive, they probably would look different from humans of today. They probably would have squat bodies with thick legs and large ears.

(b). Effect on earth's Axial Tilt (also Known as Obliquity) and Seasons

Without the moon, there might not be any seasons on the earth, or the seasons might be very different ones.

Earth has an imaginary axis going right through the canter of the earth from top to bottom. earth spins around this axis, making one complete turn each day. That is why there are days and nights on earth. The axis is not exactly vertical to its orbital plane around the sun. It is slightly tilted, and the tilt keeps varying over a long period of time approximately between 22 and 24 degrees. The cycle will be completed in about 41,000 years. At present, the earth is heading towards its minimum tilt in the next 13000 years.

The earth might have originally started off with almost a perpendicular rotation axis. But for the impact force of Theia to form the moon tilted earth and the tilt causes seasons. The tilted axis always points in the same direction. So, when the north pole tilts towards the sun in June, it heats up the northern hemisphere more directly and its summer there. Six months later, in December, the south pole warms up with its tilt towards the sun. So, throughout the year, summer and winter keep exchanging their places between the two hemispheres, all because of that tilt of the earth. The equator receives abundant sun throughout the year, and it is warm all the time.

> **Note:** With a near-zero-degree tilt, earth would have barely a few minutes of sun; or like Uranus, with its 97-degree tilt, earth would have 42 continuous years of sunlight, followed by 42 years of darkness.

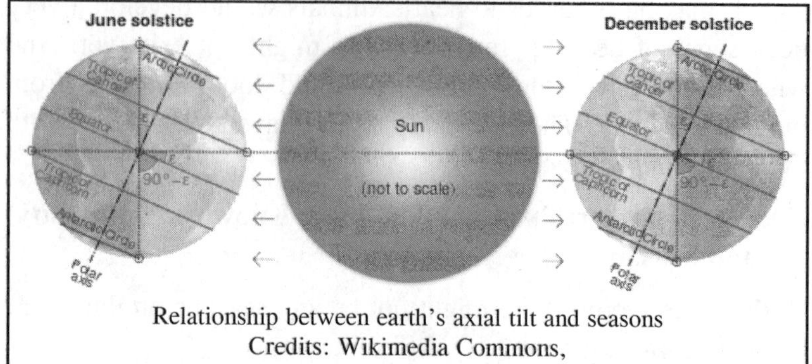

Relationship between earth's axial tilt and seasons
Credits: Wikimedia Commons,
https://commons.wikimedia.org/wiki/File:Axial_tilt_vs_tropical_and_polar_circles.svg
Author: cmglee, NASA. July 1, 2015 (NASA bitmap: July 1, 2004)
Figure in Public Domain

(c). Stabilization of earth's Wobbling (also Known as Nutation)

As the earth rotates, it also wobbles slightly back and forth on its axis. It is like a top, which doesn't simply spin in a vertical position on a floor. This wobbling is caused by earth's surface matter being redistributed. To make earth wobble, large amounts of mass need to be moved from one place to another so that earth is "off balance." The North Pole then adjusts to a new position to compensate. For example, large amounts of water are displaced seasonally when glaciers and ice sheets melt in spring. The mass shifts back when they refreeze in winter. The moon stabilizes earth's rotation. Gravitational attraction between earth and the moon is responsible for this stabilization. Without the moon, earth's axis would have continued to wobble, like moonless planet Mars. Thus, the moon keeps earth from wobbling violently as it spins and keeps earth's climate fairly controlled.

(d). Earth's Atmosphere Would Have Higher Carbon Dioxide and Lower Oxygen Concentrations

The early atmosphere of earth was made up mainly of carbon dioxide, nitrogen, and water vapor. The air was also much denser than it is now. At the time of Theia impact, part of that atmosphere also got ripped away and the atmosphere became thinner. Eventually, the first tiny plants appeared in the seas. They used carbon dioxide as food and gave off oxygen as a waste material (photosynthesis). Oxygen level slowly increased in the atmosphere. Without the Theia impact there would have been no moon in the sky, carbon dioxide would have remained higher in the atmosphere and would have made earth a much hotter place. Atmospheric oxygen level would have grown much slower and many of the plants and animals would have been different.

(e). There Would be No Eclipses on earth

The sun shines on the earth, and the earth casts its long shadow for over a million kilometres in its wake in the empty space. It is only the moon which occasionally passes through the shadow to produce the grand spectacle of a lunar eclipse. There would also be no solar eclipses either. The moon's shadow is almost equal in length to the earth-moon distance; without the moon, there will be no shadow, and no disc to block the sun.

(f). Nights Would be Much, Much Darker

You have pretty decent night vision, so long as the moon is out in the sky. But without the moon, your night vision will not be very effective. It is probably safe to say that vision would have evolved somewhat differently without the moon and that your nights would provide you with a wildly different world to experience.

Reference

Comins, Neil F. What if the Moon Did't Exist? Voyages to Earths That Might Have Been. HarperCollins. November 1993.

3. Health Issues

Q267. How does space/lunar exploration affect human health?

A267. Several consistent medical problems have been encountered by astronauts during space flights. Most of these can be attributed to the absence of or low gravitational force, low atmospheric pressure (hypobaric) syndrome, and exposure to radiations.

Some of the medical problems are vestibular dysfunction, muscle mass loss, increase in height, upward fluid shift, anemia, cardiovascular deconditioning, muscle atrophy, bone loss, and blurriness and impaired eyesight. While some of them are adaptive in nature and could be reversible, the most disturbing of all these problems is relentless bone loss. Bone loss occurs in the weightless environment of space because bones no longer have to support the body against gravity. The bone loss creates negative calcium balance. This problem appears to be irreversible, and critical demineralization can occur after a long duration in a weightless state.

Scientists think the fluid redistribution problem due to micro-gravity leads to nasal congestion, puffy faces, eyesight issues, and red blood cell death. Scientists have found a simple solution for upward fluid shift, due to micro-gravity, causing eye problems. Scientists have proposed a suctioning sleeping bag high that unloads pressure in the brain by suctioning fluids into the lower body. NASA hopes the sack can be used by astronauts in space to alleviate the vision problems they commonly endure during longer missions.

Problems due to low atmospheric pressure (hypobaric) syndromes are cardiovascular disease, respiratory disease.

Talking about the lunar environment, exposures to very low atmospheric pressure, low gravity, space radiation, and lunar dust represent the gravest threats to human health. The risk of hypobaric injury from the lack of atmosphere is a widespread phenomenon for astronauts living and working on the moon. Deconditioning of the cardiovascular system in microgravity is a well-known hazard. Without appropriate protection, the radiation environment on the moon can produce an array of both acute and chronic health problems. Finally, lunar dust, as a highly fragmented and highly reactive substance, has the potential to cause respiratory illness, which may be exacerbated by differential patterns of particle deposition in the lungs under hypo- gravity conditions.

References

1. *Robert J. Reynolds (February 20, 2019). Human Health in the Lunar Environment, Lunar Science, Yann H. Chemin, IntechOpen, DOI: 10.5772/intechopen.84352.*
   268
   https://www.intechopen.com/chapters/65461
2. *Tavassoli, M. Medical problems of space flight. National Center for Biotechnology Information, National Library of Medicines.*
   PMID: 3490786 DOI: 10.1016/0002-9343(86)90357-8
   https://pubmed.ncbi.nlm.nih.gov/3490786/
3. *Effect of spaceflight on the human body*
   From Wikipedia, the free encyclopedia
   https://en.wikipedia.org/wiki/Effect_of_spaceflight_on_the_human_body

4. Moon Landings Hoax

Q268. Why do some people still think that the moon landings were hoax and what are NASA's defense arguments?

A268. It all started with a man called Bill Kaysing and his self-published pamphlet in 1976 entitled 'America's $30bn swindle' and his book titled 'We Never Went to the moon: America's Thirty Billion Dollar Swindle', which was self-published in 1976 and republished by Health Research Books in 2002. It is interesting to note that even now some people do not believe that the moon landing really happened. According to a Voice of America article of July 20, 2019, "Despite the huge amount of evidence, the dust and rock samples, the television footage, and the hundreds of thousands of people who made it happen, polls show as many as 6 per cent of Americans believe the Apollo 11 astronauts never landed on the moon." Conspiracy theorists continue to insist that the entire mission 50 years ago was an elaborate hoax, produced at the Area 51 Air Force testing range in Nevada or on a Hollywood movie soundstage by legendary director Stanley Kubrick (producer and director of science fiction epic 2001: A Space Odyssey in 1968).

Some of the arguments mentioned in Kaysing's book and raised by some others are:

1. NASA lacked the technical expertise to put a man on the moon.
2. The absence of stars and some other unexplained optical anomalies in lunar surface photographs was indicative of a hoax.
3. There was an absence of blast craters beneath the Lunar Modules and the rocket engines of the Lunar Modules (LM) should have generated an enormous dust cloud near their landing sites in the final seconds of descent.
4. The death of Thomas Baron, a quality control and safety inspector for North American Aviation, was mysterious and indicative of a hoax.
5. There is no wind on the moon, but the American flag appears to wave in the lunar wind.
6. Multiple-angle shadows in the moon photos prove there was more than one source of light, like a large studio lamp. Also, the moon's temperature is toasty 140 degrees C (280 degrees F) in the sun's rays. The photo film (among other things) would have melted.
7. To leave a footprint requires moisture in the soil, doesn't it?

NASA's Defence Arguments

NASA has done appreciable debunking work over the decades. NASA spokesperson Allard Beutel (news chief in Public Affairs at NASA's Kennedy Space Center in Florida) recited a pile of evidence supporting the moon landings. He has mentioned that the returned moon rocks, the ability to bounce laser beams off gear the astronauts left behind, and images NASA's Lunar Reconnaissance Orbiter (LRO) took of the Apollo landing sites in 2011 are ample proofs confirming lunar landings by the astronauts.

NASA Facts published in June 1977 and reissued on February 14, 2001, has also answered the question which has been asked to them repeatedly by some "DID US ASTRONAUTS REALLY LAND ON THE MOON? Salient points from the publication are summarized below:

1. Yes. Astronauts did land on the moon.
2. Apart from the fact that millions of people saw the Apollo series on television and heard them on radio in real time, perhaps the lunar material is as irrefutable proof as any that the moon missions were not "faked." The rocks and particles still under study by scientists worldwide show major chemical differences from any previously known earth rocks. This material could not have been collected, or even manufactured on earth, and clearly comes from an entirely different world.
3. Since its inception in 1958, NASA has operated an "open" programme, i.e., all activities have been covered in depth by the news media. For example, during Apollo 11, over 3,500 media representatives from all over the world were at Kennedy Space Center, Florida, for the liftoff. Most of these press, television and radio reporters shifted immediately after launch to Johnson Space Center, Texas, to follow the operational phases through splashdown in the Pacific Ocean and recovery by an aircraft carrier with the US President aboard. Before undertaking the Apollo programme, NASA had to justify it to the President and Congress before funds were appropriated.
4. The author of the book We Never Went To The Moon, alleges that there is no visible crater (produced by engine blast) under the lunar module (LM) in photographs of it

resting on the moon's surface. His ideas in this respect seem to be based on two misunderstandings about the moon:

(i) Although the descent engine on the LM is powerful, most of its operation takes place thousands of feet above the moon during the early stages of the landing. Because the lunar atmosphere is a near-perfect vacuum, no air currents are set up that would scour the surface at a distance as might occur during a landing on earth. Movies of the landing show that, at the moment of touchdown, a small amount of surface dust is blown away, but the relatively cohesive lunar surface seems to deflect the blast sideways, rather than developing a crater directly underneath the LM.

(ii) The lunar soil is not a fluffy dust but a moderately dense and cohesive material somewhat like wet sand or ploughed farm soil. Therefore, it is not surprising that the LM engine did not excavate a crater. The lunar surface, in fact, turned out to be much more dense, compact, and resistant to penetration than some scientists originally thought. All evidence indicate that the lunar soil is too cohesive for the LM descent engine to excavate a large crater during the lunar landing.

Following points are also given is support of the moon landings:

Flag: the flag's edges are pulled tight. This effect, which was done purposely as to not allow the flag to just hang flat, it was created by inserting a stiff wire into the fabric. The "flutter" was created as the astronauts worked to erect the flag. As the wire was adjusted, the flag appeared to wave.

NASA itself addressed the hoax accusations, saying: “Not every flag needs a breeze – at least not in space.....when astronauts were planting the flagpole, they rotated it back and forth to better penetrate the lunar soil......So of course, the flag waved. Unfurling a piece of rolled-up cloth will naturally result in waves and ripples – no breeze required......In addition to the vertical pole that supported the side of the flag, they included a horizontal arm along the top of the flag to hold it out. Sometimes the astronauts did not fully extend that arm, so the flag would not be stretched flat. That would leave some ripples in the fabric, making the flag appear in photographs as if it were waving in the breeze, even when it was perfectly motionless."

Many Lights: The astronauts were taking their photos on a hilly, brightly lit landscape while the sun was close to the horizon. Imagine taking a photograph of someone on a rolling, uneven field of snow during a full, low-hanging moon. The contours of the ground would produce shadows of many different lengths.

Stars: If you've ever taken a photograph outside at night, you'll notice that faint distant objects don't show up. It is because the brightness of the nearby objects washes out the film. In fact, if you were standing on the day side of the moon, you'd have to somehow block the landscape out in order for your eyes to adapt enough to pick out the stars.

Note: My personal opinion on this issue is what Art Harmon (a former legislative director for the US House of Representatives and currently leading the Coalition to Save Manned Space Exploration) once said "They're just troublemakers. There's always people who will say 'this never happened' or 'that never happened.' They're just trying to divide people. We went (to the moon)," And also what Del de la Haye (Film and AVID video Editor/ Cameraman/Photographer 1967-2019) has said in Quora about those who still believe that the moon landing is a hoax "Everything they have come up with has easily been debunked by a minimal amount of knowledge, of physics, photography, video, communications, science in general and common sense".

References

1. *Bill Kaysing*
From Wikipedia, the free encyclopedia
https://en.wikipedia.org/wiki/Bill_Kaysing

2. *Schwartz, Ken. Millions Still Believe the 1969 Moon Landing Was a Hoax. VOA news (Voice of America). July 20, 2019.*
https://www.voanews.com/usa/millions-still-believe-1969-moon-landing-was-hoax
3. *NASA Facts February 14, 2001 (Reissued).*
DID US ASTRONAUTS REALLY LAND ON THE MOON
https://history.nasa.gov/PDFs/lunar_landing.pdf

5. Moon's Effects on Life on earth

Q269. How does the moon affect life on earth?

A269. The moon has played a pivotal role in the development of life on earth. It shapes the course of our life in three main ways: time (lunar clock), tides, and light. The moon is also helping in maintaining earth's magnetic field.

Lunar clock

Life is regulated by both solar and lunar cycles. Solar cycle regulates "day and night cycles" (also known as "sleep–wake cycle") and repeats roughly every 24 hours and driven by earth orbiting the sun. "Lunar clock" have effects on different types of organisms. Some animals respond to both day and night cycles and a lunar clock.

Some examples are:

1. Some water animals such as crab, lobster, shrimp, or barnacle, known as crustaceans, have two compasses, a sun compass in their brain and a moon compass in their antennae. So, they know to stay buried during the day and can forage efficiently during the night at low tide.
2. The mass spawning of corals on the Great Barrier Reef is also triggered by the moon. Scientists believe an interaction between sea temperature, sun's intensity, and the phase of the moon triggers this synchronized spawning.

3. Barau's petrel, an endangered seabird, breeds on the island of Reunion in the Indian Ocean. The petrels time their pre- and post-breeding migrations by length of the day. While their arrival dates at the breeding grounds vary each year, they always arrive together at the full moon. This suggests they use both the moon and the length of the day to synchronize their migrations.

Tides

Tides are the most observable effect of the moon on life on earth. The animals that live in high tidal areas have adapted to some of the most variable and extreme places on earth. For instance, the life of a limpet (a group of aquatic snails) is quite a difficult one. One minute they are above water exposed to shoreline predators, wind, and heat - the next minute they are entirely submerged in water by waves. It has taken some remarkable adaptations for them to survive in their changeable seashore habitats.

Moonlight

Some organisms rely directly on moonlight to navigate and reproduce. Some of the examples are:

1. In bright moonlight, Ephedra foeminea cones (a non-flowering joint pine) secretes small translucent globules of sugary liquid to help stick its pollen to passing insects. These droplets sparkle under moonlight, creating a glittering display that draws beetles and other nocturnal insects to the plant.
2. Some dung beetles orient themselves using the moonlight polarization pattern and the way it scatters through the atmosphere to navigate in a straight line. They cannot travel in a straight line under non-polarizing light.

Protecting earth's magnetic field

The earth's magnetic field protects us from the sun's radiation. Earth's magnetic field is produced by the rapid motion of liquid iron alloy in the earth's outer core (geodynamo effect). According to the National Centre for Scientific Research (CNRS), the moon is thought to have kept the geodynamo active by not letting the earth core cool down.

References

1. *Lotzof, Kerry. How does the moon affect life on earth? From https://www.nhm.ac.uk/. © The Trustees of the Natural History Museum, London. Re-used under the terms of the Open Government License: https://www.nationalarchives.gov.uk/doc/open-government-licence/version/3/.*
   https://www.nhm.ac.uk/discover/how-does-the-moon-affect-life-on-earth.html
2. *CNRS. "The moon may play a major role in maintaining earth's magnetic field." ScienceDaily. ScienceDaily, April 1, 2016.*
   <www.sciencedaily.com/releases/2016/04/160401075118.htm>

6. Moon in Ancient Civilizations/cultures

Q270. Why the moon is so important in many cultures?

A270.

> **Note:** This topic is so vast that a complete book can be written on this subject. Glimpses of seven ancient civilizations are covered here: Aboriginal Australia, Mesopotamia, Maya, Ancient India, Egypt, China, and Greece. Each civilization has more than one story. One from each civilization has been written here. Readers who wish to read more can refer to some of the references given here.

We are all aware that during the day, the sun is the most luminous object in the sky and at night, it is the moon (when it is visible). For most of history, the moon was people's only night-time source of light. It was used to help travelers navigate and allowed people to work until the invention of candles, gas lamps and more recently electric lighting. This dichotomy of night and day, darkness and light, are why the moon and the sun were often worshiped by ancient cultures. The moon has long been associated with fantasy and mythology and throughout history fascinating legends (traditional stories) and myths (supernatural and sacred stories) are associated with the moon and its cycles.

Look at some of the moon's myths and legends to get a better sense of how different people see its beauty and wonder. The folktale of a "rabbit in the moon" is popular in several countries. A fox, monkey, and rabbit resolve to help a hungry old beggar on a full-moon day. The monkey gathers tree nuts, the fox steals milk. But the rabbit, who can only gather grass, offers his own body, throwing himself into the man's hearth. He doesn't burn. The beggar is a god in disguise, who rewards the rabbit by etching the act on the moon for all to see.

An Aborigine's View of the Moon

In most Aboriginal cultures, the sun is a woman, and the moon is a man. The sun woman pursuing the moon man across the sky from day to day, occasionally meeting during an eclipse.

The Yolngu people (an aggregation of Aboriginal Australian people inhabiting in the Northern Territory of Australia) call the moon Ngalindi who too travels across the sky. Originally, he was a fat lazy man (corresponding to the full moon). He drowns his children for which his wives punished him by chopping him a bit producing the waning moon. He escaped by floating up into the sky to follow the sun but could not survive due to wounds and died (the new moon). He rose again after 3 days growing round and fat (the waxing moon). The wives attacked him again after two weeks. The cycle continues to repeat every month. Until Ngalindi first died, everyone on earth was immortal, but he cursed humans and animals so that only he could return to life. For everyone else, death would thereafter be final.

Mesopotamian

The moon played a key role in Mesopotamian religious culture. As it moved through its phases, people learned to keep their calendars based on the lunar month. Nanna, also called Sin (or Suen) was the god of the moon (the son of the sky god Enlil and the grain goddess Ninlil) who played a longstanding role in Mesopotamian religion and mythology. His wife was the reed goddess Ningal ("Great Lady"), who bore him Shamash (Sumerian: Utu, "sun") and Ishtar (Inanna), the goddess of love and war.

The moon god is called by different names according to various phases of the moon. Sin was associated with the crescent moon, while Nanna was connected either to the full or the new moon. The horns of a bull were also sometimes equated with the moon's crescent.

Nanna bestowed fertility and prosperity on the cowherds, governing the rise of the waters, the growth of reeds, the increase of the herd, and therefore the quantity of dairy products produced. The worship of Nanna was associated with the breeding of cattle, which was a key part of the economy of the lower Euphrates valley. Each spring, Nanna's worshipers reenacted his mythological visit to his father, Enlil, at Nippur with a ritual journey, carrying with them the first dairy products of the year.

Nanna's mythological story is regarding the love of his father and mother. Enlil saw the virgin Ninlil bathing in a sacred river and fell in love with her and seduced her. The assembly of the gods punished Enlil for this transgression and sent him to the underworld. Ninlil, knowing she was pregnant, followed Enlil to the world of the dead, determined that "my master's seed can go up to the heavens!" Once the moon god was born in the underworld, three additional deities were born to his parents, allowing Nanna/Suen to take his place in the skies to light up the night.

Gradually Nanna became more human, being represented as a cowherd or boatman, an old man with a flowing beard, a wise god, wearing a headdress of four horns surmounted by a crescent moon. The last king of Babylon, Nabonidus (reigned c. 556–539 BC), attempted to elevate Sin to a supreme position within the pantheon.

In art, Nanna's symbols are the crescent moon, the bull, and the tripod. The moon god, Nanna (or Suen/Sin) was sometimes pictured as riding his crescent moon-boat as it made its monthly journey through the skies.

original 1915 caption: "Worship of the Moon God. Cylinder-seal of Khashkhamer, patesi of Ishkun-Sin (in North Babylonia), and vassal of Ur-Engur, king of Ur (c. 2400 BC) (British Museum). Photo: Mansell"
Credits: Wikimedia Commons
https://commons.wikimedia.org/wiki/File:Khashkhamer_seal_moon_worship.jpg
Source: Donald A. Mackenzie, Myths of Babylonia and Assyria (1915), p. 50 Messrs. Mansell & Co.
Author: Unknown
Date published 1915
Figure in Public Domain

Maya Moon goddess

Mayans generally assume the moon to be a female and the moon's phases are conceived as the stages of a woman's life. She is associated with sexuality and procreation, fertility, and growth for human beings, vegetation, and crops. The moon goddess is also a goddess of disease. She is also associated with water.

The sun was of utmost importance to the ancient Maya. The moon was nearly as important as the sun to the ancient Maya. Mayan dynasties often claimed to be descended from the moon. Mayan mythology generally associated the moon with a maiden, an old woman, and/or a rabbit. The primary Maya moon goddess was IxChel, a powerful goddess who battled with the sun and made him descend into the underworld every night. Although she was a fearsome goddess, she was also the patroness of childbirth and fertility. In Classic Maya art, the moon Goddess is shown as a young woman holding her rabbit, and framed by the crescent of the waxing moon, which is her most important, identifying attribute.

Goddess Ixchel with rabbit
Credits: Wikimedia Commons
https://commons.wikimedia.org/wiki/File:Goddess_O_Ixchel.jpg
Source: Museum of Fine Arts Boston MA 1988.1282
Author: unknown Maya artist
Date: Classic period
Figure in Public domain

India

In India, the moon is a charming romantic god called Chandra who constantly flirted with the stars. He is also called Soma. He rides through the sky in a chariot drawn by white horses. Soma is also the name of a drink said to be consumed by the Gods. In art, Chandra is shown riding a chariot.

In Vedic times, the sky was divided into 28 parts. Twenty-seven of these were occupied by stars who were goddesses known as Nakshatras. As per the Hindu Mythology, the 27 Nakshtras in the sky are the daughters of King Dakshan. The moon had married all the 27. But he loved to spend time with only one of them, his most beloved queen Rohini Nakshtara. Other wives (Nakshtras) complained to their father. But the moon did not change even after repeated requests by the King. That made the King angry. He cursed the moon, causing it to decrease in size. The moon started waning. All Devas intervened and requested the King Dakshan to withdraw the curse. The King was given assurance that the moon would start visiting all his wives. Since the curse cannot be revoked completely, therefore, the King said the moon will regain his strength for half a month. That is the reason the moon transits all 27 Nakshtras in a month, completing the zodiac orbit causing Amavasya and Poornima by waxing and waning in size. On new moon nights, when he is not seen, he was with no wife.

Chandra, The Moon God (Folio from a Book of Dreams)
Description: India, Rajasthan, Mewar, Udaipur
Drawings in watercolors Opaque watercolor, gold, and ink on paper
Gift of Paul F. Walter (M.83.219.2)
Collection: South and Southeast Asian Art
Date between 1700 and 1725
Collection: Los Angeles County Museum of Art
Taken from: Wikimedia Commons
https://commons.wikimedia.org/wiki/File:Chandra,_The_moon_God;_Folio_from_a_Book_of_Dreams_LACMA_M.83.219.2_(2_of_3).jpg
Figure in Public Domain

Egyptian God of the Moon

The moon has always played an important role in Egyptian religion. It was considered by them to be the nightly replacement of the sun.

Khonsu is the ancient Egyptian god of the moon. Mut was his mother and Amun his father. He was instrumental in the creation of new life in all living creatures. He was also considered as a "ruler of the stars".

Khonsu's name means "traveler". He is frequently shown traversing the sky in a boat. He was thought to watch over those who travel at night. As the god of light in the night, Khonsu was invoked to protect against wild animals, and aid with healing. It was said that when Khonsu caused the crescent moon to shine, women conceived, cattle became fertile, and all nostrils and every throat was filled with fresh air. The ancient Egyptians believed that sowing was best done at the time of a full moon. It was even thought that minerals in the desert came into being under the moon's influence.

In art, Khonsu is typically depicted as a mummy with the symbol of childhood, a sidelock of hair, as well as the menat necklace with crook and flail. It is most commonly represented as a combination of the full-moon disk with the crescent moon and were almost always shown with this symbol on their heads. On account of the similarity in shape of the crescent moon and a bull's horns, lunar gods are frequently described with "sharp horns".

Egyptian Gods
Figure in public domain
https://www.publicdomainpictures.net/en/free-download.php?image=egyptian-gods&id=48824

China

Chang’e is the Chinese goddess of the moon. Chang'e's pet rabbit, is the Jade Rabbit. She is the subject of several legends in Chinese mythology. In one version, in a very distant past, Chang'e was a beautiful woman. Ten suns had risen together into the skies and scorched the earth, thus causing hardship for the people. Chang’e’s husband, Hou yi, the archer (a mythological Chinese archer, a god of archery descended from heaven to aid mankind) shot down nine of them, leaving just one sun, and was given elixirs (enough for two) of immortality as a reward. While Hou yi went out hunting, Chang'e stole the elixir and drank it and flew to the moon so that her husband could not go after her. She is accompanied by a small group of moon rabbits.

In Chinese tradition, the full moon is a symbol of peace, prosperity, and family reunion. The round shape of the moon during moon festival, also known as Mid-Autumn Festival (the 15th day of the 8th month of the lunar calendar) symbolizes family reunion. During the festival an open-air altar is set up facing the moon for the worship of Chang'e. New pastries are put on the altar for her to bless, and she is said to endow her worshippers with beauty. Her story is commonly used as a cautionary tale from older generations to warn young girls about the dangers of following selfish desires.

The Moon Goddess Chang’e
Credits: Wikimedia Commons
https://commons.wikimedia.org/wiki/File:The_moon_Goddess_Chang_E_-_Unidentified_artist,_after_Tang_Yin.jpg
Artist: Unknown artist, after Tang Yin (1470–1524)
Date: Ming dynasty (1368–1644)
Place of creation: China
Credit line: Gift of Douglas Dillon, 1981
Source/Photographer: The Moon Goddess Chang’e. New York: The Metropolitan Museum of Art.
Figure in public domain

Selene/Luna (Greek and Roman)

According to the ancient Greeks, Selene was the goddess of the moon. Selene was linked to Artemis as well as Hecate; all three were considered lunar goddesses.

In the myths associated with Selene, she was a daughter of Hyperion (Greek god of light) and Theia (Greek goddess of sight). Her siblings were Helios (the sun) and Eos (the dawn). Helios and Selene were responsible for controlling the movements of the sun and the moon across the sky. Selene drove the white chariot of the moon across the sky each night. Her brother, Helios, did the same thing each morning in his golden sun chariot. While the sun chariot has four horses, Selene's usually has two. Selene had a Roman mythological equivalent, the goddess Luna.

Selene was traditionally depicted as a beautiful young woman. She often wore a crown with a crescent moon. Several lovers are attributed to her in various myths, including Zeus, Pan, and the mortal Endymion. She is best known for her affair with the beautiful mortal Endymion.

In one version of the Endymion myth, one night, while moving across the night sky, she (Selene) looked down to the earth. She saw a beautiful young man, Endymion sleeping. Endymion was a shepherd tending his sheep in the countryside Selene asked Zeus (the leader of the Greek gods) to put Endymion in an eternal sleep so that he can remain ageless and deathless. Zeus agreed and Endymion was placed into a sleep for all time. Thereafter, Endymion slept in a cave on Mount Latmos. Selene visited the cave every night. Endymion slept with his eyes open, so that he too could gaze upon his lover. Selene gave birth to 50 daughters (50 goddesses of the lunar months) as a result of her visits to Endymion. Their daughters represented the 50 lunar months of the Olympiad, or period of four years marking the beginning of the Olympic Games in ancient Greece.

Selene and Endymion
Credits: Taken from Wikimedia Commons
https://commons.wikimedia.org/wiki/File:Sebastiano_Ricci_015.jpg
Artist: Sebastiano Ricci (1659–1734)
Object type Oil painting on canvas
Date: circa 1713
Collection: Chiswick House, Current location: London
Source/Photographer: The Yorck Project (2002) 10.000 Meisterwerke der Malerei (DVD-ROM), distributed by DIRECTMEDIA Publishing GmbH. ISBN: 3936122202.
Figure in Public Domain

References

1. *Vercillo, Kathryn. 10 Interesting Myths and Legends About the moon. Exemplore. July 3, 2020.* https://exemplore.com/misc/10-Interesting-Myths-and-Legends-About-The-moon
2. *Greek Legends and Myths* *THE TITAN SELENE IN GREEK MYTHOLOGY* https://www.greeklegendsandmyths.com/selene.html
3. *Dunn, Jimmy (writing as Jefferson Monet). The Moon in Ancient Egypt. Tour Egypt.* http://www.touregypt.net/featurestories/moon.htm
4. *Myths and Legends-Moon. Myths Encyclopedia.* http://www.mythencyclopedia.com/Mi-Ni/moon.html#ixzz7HLmI
5. *List of lunar deities* *From Wikipedia, the free encyclopedia* https://en.wikipedia.org/wiki/List_of_lunar_deities
6. *Myths About Moon. Windows to the Universe.* https://windows2universe.org/mythology/planets/earth/moon.html

7. Russia-Ukraine War: Impacts on Space Programme

Q271. Will Russia's war in Ukraine hinder space research and exploration?

A271. Note: Russia's war in Ukraine was going on at the time of writing this book.

Russia's ongoing invasion of Ukraine, and subsequent sanctions by European Union, United States and others may have a far-reaching consequence in space exploration activities because Russia is involved in many projects in collaboration with other countries.

A wide range of space missions is facing postponements or cancelations due to escalating tensions on the ground in the wake of Russia's full-scale invasion into Ukraine on February 24, 2022.

It is a sad and uncertain time as we watch the battle for Ukraine unfold.

Here's a closer look at some of those projects.

Impact on the International Space Station (ISS)

International Space Station (ISS) has continuously hosted crew from the US, Russia, Europe, and Japan since November 2000. The current crew (March 2022) includes one astronaut from the European Space Agency (ESA), four from NASA, and two from Russia's Roscosmos.

Russia has hinted that it would not extend its participation to ISS past 2024 in response to sanctions. Russia and China already have plans for a proposed joint International Lunar Research Station in the early 2030s. The Russian Progress vehicles provide the periodic boosts the station needs to keep its orbit. If Russia decides to discontinue participation in ISS, then NASA and others must explore the possibility of continuing station operations without Russian assistance.

Impact on interplanetary missions

The Russia-Ukraine war is also affecting planetary exploration and future launches. Already, the European Space Agency (ESA) has announced that the ExoMars Rosalind Franklin rover, which would use a Russian launch vehicle and lander, won't launch this year.

Roscosmos has also tweeted that it won't collaborate with NASA on a future Venera-D mission to Venus.

Impacts on launches

In response to sanctions, Roscosmos is pulling out of Soyuz rocket launches from ESA's Kourou Space Centre in French Guiana. Russian personnel are withdrawing as well. This change could delay the Euclid Space Telescope launch, currently still set for 2023.Several other launches which have been planned using Russian Soyuz rockets will be searching alternates, hence may get delayed.

ExoMars rover

The ExoMars mission is a partnership between the European Space Agency and the Russian space agency Roscosmos. The orbiter has been in Mars orbit since late 2016, but the Rosalind Franklin rover was supposed to launch in September 2022. Now it's at a risk of slipping to 2024.

References

1. *Dickinson, David. Russia-Ukraine war: Impacts on space. Sky & telescopes. March 3, 2022.* https://skyandtelescope.org/astronomy-blogs/astronomy-space-david-dickinson/russia-ukraine-war-impacts-on-space/?utm_source=cc&utm_medium=newsletter
2. *Kruesi, Liz. How Russia's war in Ukraine hinders space research and exploration. Science News. March 3, 2022.* https://www.sciencenews.org/article/russia-ukraine-war-space-science-nasa

3. *Davenport, Christian. Russia cuts off rocket engine supply and threatens space station partnership. The Washington Post. March 3, 2022.*
https://www.washingtonpost.com/technology/2022/03/03/russia-nasa-rocket-engines-rogozin-ukraine/

❑